# Loganair

## A Scottish Survivor
### 1962-2012

# Loganair

## A Scottish Survivor
## 1962-2012

## by Scott Grier

Published for Loganair Limited by
**kea publishing**
www.keapublishing.com

Published in 2012 for Loganair Limited by
**kea publishing**
14 Flures Crescent
Erskine
Renfrewshire PA8 7DJ
Scotland

**British Library Cataloguing in Publication Data**

Loganair: A Scottish Survivor, 1962-2012.
A catalogue record for this book is available on request from the British Library
ISBN-13: 978 0 9564477 2 2

Typeset in Scotland by Delta Mac Artwork, Email: deltamacartwork@btinternet.com

Printed in Scotland by Kestrel Press (Irvine) Ltd, 25 Whittle Place, Irvine, Ayrshire KA11 4HR

*Loganair: A Scottish Survivor, 1962-2012* is dedicated to the several thousand Loganair staff who, from the early years of Duncan McIntosh, Ken Foster and Gilbert Fraser, right through to today, have worked so hard to provide safe and reliable air services in Scotland and beyond.

# Contents

# Abbreviations

| | |
|---|---|
| AA | American Airlines |
| AAIB | Air Accident Investigation Branch |
| ABH | Airlines of Britain Holdings Ltd |
| ACMI | Aircraft, Crew, Maintenance and Insurance |
| ADS | Air Discount Scheme |
| AFC | Air Force Cross |
| AOC | Air Operator's Certificate |
| ATLB | Air Transport Licensing Board |
| ATP | Advanced Turbo Prop |
| BAA | British Airports Authority |
| BACX | British Airways CitiExpress |
| BAe | British Aerospace |
| BAHD | British Airways Highland Division |
| BALPA | British Air Line Pilots Association |
| BAR | British Airways Regional |
| BAS | British Air Services |
| BBC | British Broadcasting Corporation |
| BCAL | British Caledonian Airways |
| BCCI | Bank of Credit and Commerce International |
| BEA | British European Airways |
| BIA | British Island Airways |
| BMA | British Midland Airways |
| BNOC | British National Oil Corporation |
| BOAC | British Overseas Airways Corporation |
| BP | British Petroleum |
| BRAL | British Regional Airlines Ltd |
| CAA | Civil Aviation Authority |
| CAEL | Caledonian Airborne Engineering Ltd |
| CnES | Comhairle nan Eilean Siar |
| CRS | Computer Reservations System |
| DHC | de Havilland Canada |
| GAC | GAC Shipping Ltd (Gulf Agency Company) |
| GPO | General Post Office |
| HF | High Frequency |

| | |
|---|---|
| HIAL | Highlands and Islands Airports Ltd |
| HIDB | Highlands and Islands Development Board |
| HITRANS | Highlands and Islands Transport Partnership |
| HRC | Highland Regional Council |
| IATA | International Air Transport Association |
| ICFC | Industrial and Commercial Finance Corporation |
| ILS | Instrument Landing System |
| IRA | Irish Republican Army |
| JAL | Jetstream Aircraft Ltd |
| LDOS | Lord's Day Observance Society |
| LIBOR | London Interbank Offered Rate |
| MBO | Management Buy-Out |
| MIA | Manchester International Airport |
| MMC | Monopolies and Mergers Commission |
| MOD | Ministry of Defence |
| MP | Member of Parliament |
| MSP | Member of the Scottish Parliament |
| NATO | North Atlantic Treaty Organisation |
| NDB | Non-directional beacon |
| OIC | Orkney Islands Council |
| OISC | Orkney Islands Shipping Company |
| OPMAC | Operation Military Aid to the Community |
| PSO | Public Service Obligation |
| RAF | Royal Air Force |
| SAS | Scandinavian Airlines System |
| SAS | Scottish Ambulance Service |
| SIC | Shetland Islands Council |
| SRC | Strathclyde Regional Council |
| STOL | Short take-off and landing |
| TWA | Trans World Airlines |
| UKAEA | United Kingdom Atomic Energy Authority |
| WIIC | Western Isles Islands Council [1975-1996] |

# Acknowledgements

I am hugely indebted to my Personal Assistant, Mrs Linda Doak, whose cheerful disposition, endless patience and total commitment not only ensured my manuscript was completed and published on time, but also helped to sustain us both throughout the good times and not so good times during no less than thirty-six of the Company's fifty years.

Special thanks are also due to Iain Hutchison for his help and guidance during this project and for having allowed me to draw on his vast knowledge and experience.

I am grateful to many others who in so many ways have been most helpful to me by sharing with me their Loganair experiences and memories, or providing information or photographs, or just sound advice:
Roy Bogle, Trevor Bush, Ken Foster, Gilbert Fraser, Jonathan Hinkles, David Hyde, Lloyd Cromwell Griffiths, Stuart Linklater, Sandy Matheson, Alex MacDonald, Charles Randak, Austin Reid, Ian Reid, David Ross, Audrey Sanders and Eddie Watt.

Finally I must thank my wife, Frieda and my sons Christopher and Geoffrey for their love and unstinting support and encouragement.

# Introduction

'This is your Captain speaking. My name is…' Time and time again, I would be sitting on a flight – British Airways, British Midland, Ryanair, Thomas Cook or whichever – and I would suddenly be aware of a name or a voice from the flight deck I recognised as one of Loganair's former pilots. Or I am awakened from my semi-comatose state by: 'Hello Mr Grier, remember me?' As I try to peer as surreptitiously as possible at the cabin staff member's name badge, I hear: 'I worked for Loganair fifteen years ago.' I am therefore constantly reminded of just how many people over the years have spent some time as a Loganair employee. Crews in particular have joined the Company knowing they will receive excellent training and experience which will stand them in good stead when moving to bigger airlines. However, many more staff have joined the Company and remained for many years, and I find it very gratifying that I have many colleagues who have more than thirty years service. I have never doubted that it is the outstanding contri-butions of a great many dedicated staff that has made Loganair a Company which is held in such regard and affection throughout Scotland.

No other airline in the UK has operated, uninterrupted, for fifty years under its original name. In that time, around fifty Scottish-based airlines have come and gone, and if airlines serving Scotland are included, that number rises significantly. Loganair, too, has come perilously close to extinction on several occasions. Many people have convinced me that the story of how Loganair has survived does deserve to be told. Having been privileged to be involved in the management of the Company for more than thirty-five years, I realised I was probably best placed to write it.

The Company was fortunate in the early days to be in the hands of Willie Logan, surely one of Scotland's great post-war entrepreneurs. Then it was the strength and paternalism of a Bank – the Royal Bank of Scotland – that allowed it to keep going. More particularly, it was down to the support and loyalty of the Bank's Deputy Chairman, John Burke, during a period of mounting financial losses and the growing impatience of his colleagues on the Bank Board. He ensured that the Company was sold to Michael Bishop and British Midland Airways who were seen as the safest hands available. This worked well for a time before the new owners fell out of love with their Scottish subsidiary, almost dismantling the Company before eventually agreeing to a management buy-out.

Running throughout much of Loganair's fifty years existence, influencing if not determining the fate of the Company, were the actions, if not antics of near neighbours, British European Airways and its successors. Loganair had a love-hate relationship with British Airways in all its guises from the mid 1960s right through to 2003 when it finally withdrew from its Highlands and Islands routes. Until 1993, scheduled air services in the UK were regulated, with route licences granted, latterly, by the Civil Aviation Authority. British Airways held the Scottish route licences and incurred heavy financial losses throughout the period. Successive Loganair managements convinced themselves that, for the Company to expand and survive, it had to have access to these routes, and they believed that with its lower cost base, Loganair could succeed where British Airways had failed.

For years, Loganair could make little progress. Then, in the 1970s, there was a period of collaboration when some of the smaller routes were taken over. The creation of Highland Division in 1981 frustrated the Company's efforts for a time, before talks between the two Companies resumed, but these talks were always overshadowed by the bitter rivalry between British Airways and Loganair's owners, Michael Bishop's Airlines of Britain Holdings. No one could have written the script for the next series of events, the many twists and turns affecting the fate of these routes. British Airways suddenly handed them over to Loganair's sister company, British Regional Airlines and then got them back by buying that Company. Later, in an ultimate acknowledgement that it could never make these Scottish routes viable, it handed over all its remaining routes to Loganair.

In each of its five decades, Loganair survived critical times. In the mid-1990s, I was attending a meeting in Shetland at a time when the Company's problems were very public and all too obvious. I was asked: 'Will the Company survive?' I certainly hoped so. 'Will the name Loganair survive?' Before I could answer, someone shouted out: 'Of course it will always be Loganair. My mother still shops at Lipton's.'

# Chapter 1

# The 'Logan' Years
# 1962–1968

### *In the beginning*

Fifty years is a long time, and memories are short. The question is very frequently asked 'Where did the name Loganair come from? Was there a Mr Logan?' Indeed there was, and fifty years ago Willie Logan was a household name in Scotland. Loganair was founded simply because Scottish industrialist Mr Willie Logan had to get himself quickly around the country from one of his construction sites to another. He had frequently resorted to chartering aircraft for the purpose, most regularly from a small air taxi firm called Capital Services (Aero) Ltd with bases at Turnhouse and Renfrew aerodromes for its single Piper Apache and collection of light, single-engine aircraft.

When the air taxi firm found itself in financial difficulties, William Logan decided to step in and took over the Company on 1 February 1962 and renamed it Loganair, the name which would be used by the airline for the next fifty years without interruption, making it the UK airline operating longest under the same name.

As a result of the takeover, 'Loganair' and all its operations were centred on Renfrew with a Piper Aztec as the sole fleet aircraft. William Logan, Chairman of the construction company, became Chairman of Loganair, while Messrs Martin Hill and

*Loganair's first aircraft in 1962 was the Piper Aztec and it remained the workhorse for passenger and newspaper charters until 1967. Other small aircraft included the Piper Cherokee and the Piper Tripacer which was acquired because of its suitability for aerial photography.*

Courtesy Capt. Ken Foster

7

Geoffrey Harrison were appointed joint Managing Directors. Captain Duncan McIntosh, who had formerly been Capital's Chief Pilot, now became Loganair's Chief Pilot and Manager. The scene was set.

### *Mr William Logan*

Willie Logan joined the small stonecutting business of his father, Duncan, in Muir of Ord, Ross-shire in 1932 and, by the 1960s, he had built Duncan Logan (Contractors) Limited into a major civil engineering company. The company had undertaken government defence contracts during the Second World War and expanded rapidly, the number of employees rising from 50 to 200. It was in the post-war years, however, that the Company's expansion really took off.

By 1962, Duncan Logan (Contractors) Limited had contracts all over Scotland and beyond. By his own estimation, Willie Logan was travelling some 2,000 miles a week from his home in Muir of Ord to his various construction sites. Little wonder then that he had frequently used the aircraft of Capital Services (Aero) Limited to get him around the country when he undoubtedly had got to know its Chief Pilot, Duncan McIntosh. When the air taxi firm was known to be struggling financially, it clearly made sense for Logan to acquire the company – and its one Piper Aztec, together with its Chief Pilot and a few other staff. So the 'aviation division' of Duncan Logan (Contractors) Limited was born, and on 1 February 1962 started trading as Loganair.

Being awarded the Tay Bridge contract was a breakthrough. 'The 49-year-old self-made Scots tycoon' certainly enjoyed a good press, capturing the public imagination with his aircraft dash to Dundee's Riverside Drive airstrip on 17 January 1963 to submit his successful Tay Road Bridge tender – in a snow storm and only fifteen minutes before the deadline. He beat seven other bidders including one other Scottish company, Whatlings Limited of Glasgow. Then, as now, jobs were important and Logan would be taking on 250 additional workers. Winning the contract would also be important to Logan's subcontractors, Dorman Long and Company (Bridge and Engineering) Limited of Middlesbrough, which was to supply £1 million of steel work. Dorman Long, in turn, would subcontract fabrication work on 8,000 tonnes of girders worth £750,000 to the conveniently located Caledon Shipbuilding and Engineering Company Limited in Dundee thereby assuring work for the local hard-hit platers and welders for two years. It was a good news story and the newspapers loved it.

By then, however, the company already had many other contracts under way. The Tay Road Bridge contract price was £3,823,153 but Willie Logan revealed that at the time of the Tay Bridge tender he was already working on six other projects of more than £1 million. Clearly not a man to hide his light under a bushel, he also revealed to

*Founder of Loganair on 1 February 1962, Willie Logan, one of Scotland finest post-war entrepreneurs. He was responsible for many noteworthy civil engineering contracts until his death in an air accident on 23 January 1966.*

Courtesy of Dingwall Museum

the press: 'I've a tender in for another £3,500,000 contract in Scotland, but I can't say just now what it is.'

That particular contract may not have materialised, but by then the extent and variety of Logan's building and civil engineering projects were quite remarkable: road and housing projects throughout the Highlands, hospitals, much of Cumbernauld new town, hydro-electric schemes, the pulp mill at Corpach, a NATO fuel depot, and the Hamilton Teachers Training College, as well as the reconstruction of a six-mile stretch of the A8 Edinburgh–Glasgow road at Baillieston and the Cumbernauld bypass at Castlecary.

The aviation dimension in the attendant publicity was not lost on Willie Logan who told the press that he also had 'a special plane standing by' to fly another £1 million tender to London 'to beat the 10am deadline' for part of the Ministry of Aviation £4.2 million contract to develop Abbotsinch Airport. Willie Logan clearly was enjoying having aircraft at his disposal.

Described as kindly and genial and with a ready smile, Willie Logan was popular with his workforce who respected his readiness to get involved and get his hands dirty. He won the reputation of having one of the shrewdest and most enterprising brains in the country, masterminding his business from his headquarters in Muir of Ord and routinely using one of his planes to fly him from the small airstrip on the nearby Highland Showground. Also, for his own convenience, he had a Jaguar car left for him in Dundee and a Rolls-Royce in Glasgow. Willie Logan had style.

Willie Logan, like his father, was a member of the Free Church. He told the Sunday Post, 'I am a wee free and I don't smoke or drink.' He was proud of his company's excellent record of keeping to contract schedule despite his ban on Sunday working. In fact, his Sabbatarian principles were tinged with pragmatism. There were times when work was carried out on Sundays – but 'only in an emergency, and very, very seldom.'

Ironically, it was on Sunday 12 December 1965 that Willie Logan suffered a broken collar bone and other minor injuries when he was struck by the hook of a crane. This occurred when the jib buckled while he was directing lifting operations of a crane sunk about 100 yards from the north shore of the Tay. Sadly, three men died in the accident and, according to an eye witness, Willie Logan was very lucky. Wrapped in a blanket, he was assisted ashore to an ambulanceman, but only after all waiting press men were ordered off the pier.

Only a month later, on Saturday 23 January 1966, William Logan was killed in an air accident. He was travelling from Edinburgh to Inverness to a meeting of the Inverness Building Society of which he was a Director. On that particular day, the Loganair aircraft was not quite ready, so he decided to charter a Piper Aztec from Auchterarder-based Strathallan Air Services, operating as Strathair, which had been formed the previous year. Unfortunately, Strathair was not on the list of non-scheduled service airline operators approved for the carriage of company personnel and named on the Logan Group Insurance Policy.

It is reported that the Piper Aztec, piloted by ex-Squadron Leader Peter Tunstall, Managing Director of Strathair, took off in the dark from Edinburgh without filing the required IFR flight plan and flew to overhead the Inverness Non-Directional Beacon (NDB) in cloud. His en route chart did not plainly show that the NDB was located not on the airport but on top of the 922-feet Craig Dunain. He descended over the Beacon expecting to emerge from the cloud base above the airport, but instead crashed on the tree covered hilltop. Captain Tunstall suffered a broken arm, and cuts and burns for which he required skin grafts, but he survived. Willie Logan was killed instantly and Scotland had lost a celebrated and favourite son.

The Logan Group Insurers refused to make any payment because Strathair was not one of the approved operators. Strathair had only the minimum £3,500 per person Passenger Liability Insurance required by the Warsaw Convention and this was paid to the Company. Mrs Helen Logan, Willie Logan's widow, soon had to leave her company-owned home and later initiated legal proceedings against Captain Tunstall.

Willie Logan is buried in Dingwall and his memorial is in the form of the two columns and deck of the Tay Road Bridge. The Fife Area Council of the Scottish National Party proposed that the Tay Bridge should be called 'The Logan Bridge' to perpetuate the name of the man who was 'synonymous with the character of a true Scotsman – compassionate, religious, astute and industrious'. It was not to be, but many thought this suggestion captured the public mood.

### Captain Duncan McIntosh

Loganair's new Chief Pilot and Manager, Duncan McIntosh, was an RAF pilot during the Second World War flying Spitfires from the aircraft carrier *HMS Furious* in Malta. He served with Fighter Command for five years from 1941 flying North American Mustangs, Hurricanes and Typhoons. On his demobilisation from the RAF in 1946, he resumed his pre-War Civil Service career while also helping to restart No. 603 City of Edinburgh Spitfire Squadron at Turnhouse where he continued to fly on a part-time basis. In 1951, he returned to flying as a civil test pilot for Airwork, testing the Royal Naval aircraft of the day – Vampires and Gannets. Six years later, he joined the Miles Aircraft Company at Shoreham in Sussex and worked on Gemini, Aerovan and Student aircraft until Miles was absorbed upon the formation of Beagle. He returned north in 1960 and joined Capital Services (Aero) Limited as Chief Pilot.

Duncan McIntosh, or 'Captain Mac' as he was known to most people, was an affable, inveterate pipe-smoker with a passion for flying. He was an extremely accomplished pilot and his consummate airmanship was admired by his peers. It was his enthusiasm for flying that also encouraged him to set up Glasgow Flying Club in

*Captain Duncan McIntosh. Never happier than when flying, Captain Mac completed 10,000 flying hours on a range of aircraft types between 1941 and 1980.*

Photo: Randak Design

1962. Captain Mac would often speak of the pioneering flying that Captain E E Fresson had undertaken in the 1930s and he was undoubtedly influenced by his achievements. Fresson had established air mail and passenger services from Longman aerodrome, Inverness, initially to Wick and Orkney, and later to Shetland and Stornoway. Duncan McIntosh was well aware that Fresson's local services had mostly been discontinued when the Second World War started and did not start up again when Captain Fresson left the newly formed British European Airways in 1947. There was therefore a gap which he and Loganair could fill.

Captain Mac would continue to be Managing Director of Loganair until his retirement on 31 December 1982. It was in the 1960s, however, that he was able to make his greatest contribution when he succeeded in picking up the baton of Captain Fresson and pioneered new air services in Scotland. Captain Mac loved flying and certainly led from the front. At that time, light aviation was largely uncontrolled. Flight and duty regulations were practically unheard of and no doubt he would sometimes fly in conditions which would not be deemed to be possible or wise today, but he successfully relied on his vast experience and pilot skills. Flying in Scotland's difficult weather conditions was challenging, but for Captain Mac it was always very personally rewarding although not always commercially sensible for the Company. He acknowledged candidly in an interview in 1965, 'We could not run a service like this if we did not have the civil engineering business behind us.' Later, under the ownership of the Royal Bank of Scotland, Captain Mac was advised that he had to do more managing and less flying. In the more formal, structured flying environment of the 1970s, and certainly in the Boardroom, Captain Mac was less comfortable and effective. He was an aviator first and foremost. His services to Scottish aviation, and the Air Ambulance Service in particular, were recognised in 1977 when he was awarded the OBE in the Queen's Honours List.

Even in the late 1970s, when he was flying very infrequently for Loganair, he brought the Miles Student to Glasgow in an endeavour to demonstrate its role as a cost-effective, versatile jet trainer. As late as 1978, he personally flew the Student, with his old skill and panache, for a range of potential customers including a deputation from King Hussein of Jordan. Alas to no avail – the aircraft really did belong to a slightly earlier period.

### Getting around Scotland in the 1960s
The Scotland of 1962 was 'bigger' than today as so much of the present transport infrastructure had yet to be built. In a country where geography presented severe difficulties and discomfort for travellers, even short journeys by surface travel could take many hours – as Willie Logan had been finding out. When Duncan McIntosh

*Throughout its existence, BEA dominated air services in Scotland. The diversity of its operations is epitomised by this scene at Renfrew Airport which shows a Handley Page Dart Herald, Vickers Viscount, de Havilland Heron, Vickers Vanguard and Douglas DC3 Dakota.*

came back to Scotland, with its poor transport infrastructure, in 1961, he saw an emerging role and a real commercial opportunity for air taxi and charter services which could complete a journey in a fraction of the time taken by surface transport.

These were the days before motorways of course, but with Scotland's numerous rivers and estuaries, it was the lack of bridges that was a major obstacle for car and bus travel. Neither the Tay nor the Forth road bridges had been built and the journey from Dundee to Turnhouse Airport in Edinburgh, for example, took two and three quarter hours by road and ferry, compared to fifteen or twenty minutes by air taxi. Driving from Edinburgh to Inverness routinely took five hours. North of Inverness, before the days of the Kessock, Dornoch and Cromarty bridges, the traveller had to face another long journey by road from Inverness to Wick. To the west, there was no Erskine Bridge, and to cross the Clyde west of Glasgow, the Renfrew or Erskine ferries had to be used. To the north-west, a journey very often would entail a long queue for the Ballachullish ferry, or if bus travel or a three-ton lorry were involved, or if it was after dark, it meant a seventeen-mile detour round the head of Loch Leven.

Road journeys in the early 1960s were neither quick nor easy and consequently air services on the Scottish mainland did increase gradually over the next thirty years. Ironically, fifty years on, aviation on mainland Scotland has gone full circle. With today's ever-improving road networks, there is now only one scheduled air service within Scotland that does not cross a water channel, Loganair's Edinburgh–Wick service. Formerly important air services, such as those from Glasgow and Edinburgh to Inverness, and to Aberdeen, or from Inverness to Wick, which offered such convenient transport links for many years, have been discontinued due to lack of passenger demand. Travellers on these overland journeys can now go by car, bus or train, more cheaply, just as quickly, and avoid what can be an unpleasant experience at today's airports.

Perhaps not surprisingly, in view of Willie Logan's own travel requirements, it was on mainland Scotland that Loganair initially saw its best commercial opportunity. Of course there were rail services in Scotland, but these had been contracting and the closure of a number of railway lines in the Highlands was currently under consideration following the Beeching Report, *The Reshaping of British Railways*, which was submitted to the Minister of Transport in 1963. Meanwhile the main providers of Highlands and Islands bus services were David MacBrayne Limited and Highland Omnibuses Limited, a subsidiary of the Scottish Bus Group, as well as numerous local bus operators. Losses were met directly or indirectly by the Secretary of State.

Fifty years ago, it was the shipping services which provided the lifeline to the islands on the West Coast and to Orkney and Shetland. To the West, services were largely operated by MacBrayne with financial support from the Secretary of State, while the Caledonian Steam Packet Company, a subsidiary of British Rail, operated a vehicle service between Kyle of Lochalsh and Kyleakin on the Isle of Skye. MacBrayne operated from its mainland ports of Glasgow, Oban, Mallaig, and Kyle of Lochalsh, and introduced three vehicle ferries, the *Hebrides*, the *Clansman* and *Columba* in 1964, but even these new ships had lifts to raise and lower vehicles

Courtesy of the late Captain
Ian Montgomery

*BEA Handley Page Dart Herald. The high-winged Handley Page Herald replaced the Pionairs on some of the Scottish routes. The Ministry of Supply had bought and leased three Heralds to British European Airways in 1961 for use on its Renfrew to Campbeltown and Islay routes and was the first turbo-prop aircraft to land on Islay. A Herald also operated the Renfrew-Aberdeen-Sumburgh route and a third operated Renfrew-Inverness-Wick-Kirkwall. These two aircraft crossed over for the return legs. This programme continued until the introduction, in 1966, of the Viscount aircraft.*

between the quay and the car deck, so that existing piers could be used if possible. *Loch Seaforth*, built in 1947, ran a daily service from Mallaig and Kyle of Lochalsh to Stornoway and on to Tarbert in Harris while the other cargo/passenger ship, *Lochiel*, built in 1939, ran a daily service from West Loch Tarbert to Islay, calling on Gigha, Jura and Colonsay three times a week. *Claymore* operated from Oban to Tobermory on Mull, Coll, Tiree and Barra and on to Lochboisdale on South Uist. When Duncan McIntosh returned to Scotland, and had set out to find commercial opportunities and sell the superior convenience of his air services, he had not reckoned with MacBraynes. Surface travel by ferry may not be quick, but it was what generations of islanders had been using and accepting for decades. He and Loganair would make no quick breakthrough. MacBraynes prevailed:

> *The Earth belongs unto The Lord,*
> *And all that it contains,*
> *Except the Western Islands,*
> *And they are David MacBraynes'.*

Meanwhile in the Northern Isles, it was the North of Scotland, Orkney and Shetland Shipping Company, a subsidiary of Coast Lines of Liverpool, which provided shipping services. These were without subsidy, something that caused the Highland Transport Board to question prophetically whether they could be sustained. *St Clair* operated twice a week between Aberdeen and Lerwick while *St Ninian* ran weekly from Leith and Aberdeen to Kirkwall and Lerwick with additional services by *St Magnus* in the peak summer months. There were also services on a weekly basis by *St Clement* and *St Roqueld* from Leith to Kirkwall, Stromness, and Aberdeen back to Leith. *St Ola*, supplemented by *St Clement* in the summer, operated across the Pentland Firth between Scrabster and Stromness with capacity for cars which had to be loaded and unloaded by ships' derricks. Little wonder that the Highlands and Islands Development Board recommended replacing *St Ola* with a roll-on roll-off ferry in order to improve access for buses and touring motorists, but this was still in the future. Interestingly, there was a growing awareness in the early 1960s that the Northern Isles ships may have to face growing competition from British European Airways (BEA).

British European Airways was set up in 1946. By 1963, in Scotland, its Dakotas (DC3s) had given way to Heralds, Viscounts and Vanguards, and the Rapides had been superseded by Herons. During the Second World War, a number of airfields had been constructed for defence and coastal reconnaissance and were subsequently used as civil airports. Scheduled service flying in Scotland, however, was still in its infancy. BEA's Scottish services were dependent on aircraft being made available from mainline BEA, and this meant that Scotland in general, and the Highlands and Islands in particular, were assigned aircraft which were not best suited to local services and certainly not to developing a scheduled service network. Many routes had 'through' services, starting in the south of England and finishing in Shetland. This may have made economic sense for the airline, but was far from ideal for the travelling public. Edinburgh and Renfrew (Glasgow Abbotsinch from 1966) were termini for flights from the south. Critically, most of the services north of Edinburgh and Glasgow operated only one rotation each day long after it had become very apparent that greater frequency was required. With BEA suffering heavy financial losses, its general view was that deploying any additional aircraft in Scotland would merely have exacerbated its financial position, and BEA Headquarters at Northolt would not approve. In 1965, Duncan McIntosh pointed out in the press that 'Anyone flying into Renfrew after 10 or 11am, and wanting to go to the islands, has missed the BEA plane and must wait until the next day.' Clearly the Scottish air traveller in the 1960s could have been better served. Local, 'feeder' services for BEA would surely help.

### *Incorporation*
When Willie Logan acquired Capital Services (Aero) Limited on 1 February 1962, the air taxi company started trading as Loganair. Following Willie Logan's death in 1966, some reorganisation of the Civil Engineering Group Companies took place, and a new company, Loganair Limited, was incorporated on 26 May 1966. Mr Martin Hill became Chairman, Duncan McIntosh became Managing Director and Thomas Low was appointed Company Secretary. Mr J C McLintock of Biggart Lumsden, Law Agents, became legal adviser and started a Loganair connection with that firm, now Biggart Baillie, which continues today. At the first meeting of Loganair Limited on 25 June 1966, which put in place banking arrangements with the National Commercial Bank of Scotland Ltd, it was agreed:

> that the Company purchase from Duncan Logan (Contractors) Limited as a going concern, the Air Charter business carried on by, and under the name 'Loganair' and all stock-in-trade, fittings and assets at a price to be agreed on valuation. As part of the Agreement for the purchase of the said business, the Company would pay the existing liabilities of, and receive payment of the debts due to the Loganair division of Duncan Logan (Contractors) Limited.

### *Where to land*
Willie Logan had found out very soon that owning a small air charter company, even one as small as Loganair, was an expensive luxury, especially as it was primarily for his own personal convenience. It was very evident to him that more revenue from third

party charter and air taxi work was required for Loganair to make better economic sense. At the same time, there was also the very strong but mistaken belief among the Loganair management that commercial opportunities abounded. One of the first jobs Duncan McIntosh had given himself upon his return to Scotland was identification of suitable landing sites for the expansion of his air taxi aircraft operations. Such was his enthusiasm that he convinced himself that as many as one hundred airstrips could be made available for commercial flying. These included at least twenty deserted ex-RAF runways, plus racecourses and private fields.

There is no doubt that some of these were unrealistic, but many were indeed developed and used for civil aviation. The Royal Engineers, under the OPMAC scheme (Operation Military Aid to the Community) built new airstrips at Plockton, Glenforsa on Mull, and Broadford on Skye. But for a last minute disagreement between the landowner and the local authority, the Army Engineers would also have built a strip at Shiskine on the island of Arran. The new towns of Cumbernauld and Glenrothes, as well as Dundee with its Riverside location, set out to establish airstrips in keeping with their status. RAF bases at Kinloss and Leuchars, and the naval air stations at Lossiemouth, Arbroath and Machrihanish were all made available for civil aviation on request. Machrihanish was already used by BEA for its scheduled service operations. On Jura, an airstrip was built by the distillery. All of these new airstrips up and down the country would surely make even the most remote parts of Scotland more accessible by air taxi. Maybe not the world, but certainly Scotland was Duncan McIntosh's oyster.

Courtesy of GAAEC

*Piper Aztec G-ASYB. The building in the background was the Company's 'headquarters' from 1966 until 1978 when new offices were erected on the same site. The hangar in the background dated from to World War Two and was only demolished in 2001 when the Company had a new hangar erected 100 yards away.*

The aviation landscape in Scotland was certainly changing. Nearer home, Loganair's own operational base at Renfrew Airport was closing. All air services were being transferred to nearby Abbotsinch, the former Royal Naval Air Station, HMS Sanderling, which had closed in 1963 and been acquired by the City of Glasgow Corporation. The day before the new Glasgow Airport was to be officially opened on 2 May 1966, Captain Ken Foster flew the Company's Cherokee 6 low over the new airport, checking the landing lights and markers on behalf of the Ministry of Aviation, then carried out the first landing. BEA made its first landing later that evening in a Dart Herald which, like Loganair's Cherokee, flew in from Renfrew. It was the end of an era and, of course, a new beginning. At 8am, on 2 May 1966, the first commercial flight to land at the new Glasgow Airport was a BEA Viscount, piloted by Captain Eric Starling, BEA's Scottish Flight Manager, carrying sixty-four members of the staff of Sir Basil Spence, Glover and Ferguson, architects of the airport. The first regular scheduled flight to land was a BEA Herald carrying forty-one fare-paying passengers from Aberdeen and Edinburgh.

### Plenty of variety

The primary reason for Willie Logan's interest in setting up Loganair was to provide quick and convenient transport for his managers and key workers between the many Logan construction sites throughout Scotland, and, of course, for himself to commute from his headquarters in Muir of Ord. Initially, the Company's only plane was a Piper Aztec based at Renfrew Airport. As early as 1963 the Company acquired a second Aztec and also a single-engined Piper Tripacer that was deemed to be particularly well-suited to aerial photography by the media and by civil engineering companies monitoring progress on their different projects. The Company was proud of the Tripacer's contribution to some spectacular photographs which appeared in the national press. On the Clyde, the atomic submarine *Dreadnought* was photographed during trials, while the activities of the United States Navy on the Holy Loch were also the subject of much public interest. It is not recorded whether the Loganair pilots had heard of the Official Secrets Act, but one assumes that Loganair's bills for these assignments were not being paid by the Kremlin! Further up the Clyde, the Cunard liner *Queen Elizabeth* was filmed as she manoeuvred into dry dock in 1965 and there were also dramatic aerial shots of the Forth Road Bridge as the magnificent suspension bridge inched its way towards completion in 1964.

The Loganair pilots were always willing to help the photographer to get the best shot, and this continued after the Britten-Norman Islanders arrived. Charles Randak, who was responsible for the Company's publicity, remembers vividly being on board one of two

Islanders to take some aerial photographs, speeding down the Kirkwall runway together, with the Control Tower shouting in disbelief and outrage that it was illegal to take off in formation and 'they were not still in the bloody RAF'. Today, of course, the two pilots would be clapped in irons. Nor would today's Health and Safety Executive be impressed at the sight of Charles Randak hanging out of the aircraft with the door removed to facilitate the camera work. Just as well he had found a piece of frayed rope on the hangar floor to tie round his waist. All in the name of art and publicity!

Hungry for work, the Loganair management had the two Aztecs employed in a whole variety of charter and air taxi work. Some of Loganair's flying exploits were beginning to capture the public imagination. Just like his boss Willie Logan, Duncan McIntosh knew how to grab the headlines. The Company, for example, was awarded a contract by the Army to take mail, food and other supplies to the radar station on the remote island of St Kilda whose native population had been evacuated in 1930. The supplies had to be dropped from the Aztec at very low level onto the main island of Hirta as there was no landing area. Loganair also assisted with radar calibration for the Army at the missile range on Benbecula. There was 'bread and butter' flying too when the Company won contracts to fly computer cards from the Post Office Savings Bank at Cowglen in Glasgow to Blackpool, and football coupons nightly for Littlewoods Pools from Renfrew to Belfast between 1963 and 1965, and from Belfast to Liverpool in 1966. The coupons contract was only lost when Littlewoods acquired its own aircraft for the purpose. There certainly was variety. The carriage of coffins – occupied or empty - became big business when the Britten-Norman Islander arrived to the extent that the Board decided to advertise in *Funeral Directors Gazette*!

The two Aztecs were still the Company's workhorses along with the Piper Tripacer. With the 'can do' approach of Duncan McIntosh and his pilot colleagues, the Company's fleet acquisition philosophy tended to be instinctive and 'horses for courses'. When Duncan Logan (Contractors) Limited won a £3 million contract for NATO, the Company thought nothing of laying out an airstrip at Aultbea for the duration of the contract, and it acquired a three-seater Piper Cherokee 180 for the operation. This was later replaced by a five-passenger Piper Cherokee Six.

The charter work was interesting and varied but, try as it might, the Company was never able to get enough productive flying hours to fully utilise its aircraft. For all the charter flying, and despite all the personal effort of Duncan McIntosh and his colleagues, the Company was not achieving an adequate financial return. When Willie Logan died in February 1966, the total annual utilisation of the two Aztecs was only 1,700 flying hours, of which 500 hours were in respect of Duncan Logan's own staff travel, and no doubt very keenly priced. Five hundred hours represented ad hoc charters where competition, real or imaginary, produced rates with wafer–thin margins. Even more telling, 700 hours, a very significant proportion, were in respect of a Stornoway newspaper contract whose daily rates seemed always to lag behind increasing operating costs,

### *Newspapers to Stornoway*

The Company thought it had made a breakthrough when, in 1964, it won the contract to carry newspapers from Glasgow to Stornoway in the Aztec. The people of Stornoway, who were now receiving their newspapers by 8.30am each day Monday to Saturday instead of, as formerly, at half past two in the afternoon when the BEA scheduled service flight arrived in Stornoway, were also pleased. It came at just the right time for Loganair and the newspaper distributors were also pleased, as well they might be, at the airline's excellent regularity. Certainly Captain McIntosh and his pilot colleagues were to earn a formidable reputation for not missing a single daily delivery, despite the often atrocious west coast weather, during the ten years the Company held the contract.

From the outset, the newspaper distributors' committee had wanted a bigger Beech 18 aircraft for the Stornoway contract as they firmly believed it was the most suitable aircraft and had the appropriate carrying capacity. Indeed, the original draft contract in 1964 had actually stipulated the Beech, but the Aztec was accepted. By late 1965 there was talk that the Newspaper Committee wanted to extend the freight flights to other outlying islands including Orkney and Shetland which of course was intended to act as encouragement for Loganair. They were particularly pressing the Company to also carry newspapers for Benbecula. To do this as well as deliver the Stornoway newspapers, which were themselves increasing in weight, the newspaper people were once again encouraging Loganair to acquire a second-hand Beech 18 aircraft, known to be available, and which had the necessary carrying capacity to do the job. Later, even after newspaper weights had increased beyond the capacity of the Beech 18 aircraft, so desperate were the newspaper people that they even promised to restrict newspaper loads to suit a Beech. They frequently requested that a Beech 18 be acquired, and, in turn, Captain McIntosh frequently recommended it to the Board, but it was 21 March 1968 before Chairman Martin Hill relented and agreed that the aircraft could be purchased if the newspaper distributors gave a Letter of Intent, guaranteeing a long term contract.

Arrangements were duly made to acquire the second-hand Beech 18, G-ASUG, which had previously been employed as an aerial survey aircraft. As it was the first Beech 18 on the UK register, the aircraft had to complete a full flight certification programme before a Certificate of Airworthiness could be granted to allow it to be used for public transport. This was carried out at Prestwick by Scottish Aviation with Captain McIntosh performing many of the test flights himself.

The Company was now operating the aircraft the customer had been wanting for some time. The Stornoway newspaper contract itself was ideal for Loganair whose main business was charter operations, providing routine, daily flights at a set time, and, of course, a regular income stream. This was surely a platform on which the management could build up the business. Certainly it gave structure and shape to the Company's overall flying operation. It also provided much needed aircraft utilisation to the extent that, by the late 1960s, it represented as much as twenty-five percent of the Company's flying hours. This would have made good business sense if only the rates had been adequate.

Unfortunately, the contract had become too big and too important to lose. The newspaper distributors were tough. Increases in rates were frequently sought by Loganair, but the newspaper distributors usually stood firm and refused, and frequently threatened that the contract would be terminated. Newspaper weights were also a constant problem. There were periods when the Company incurred heavy losses on the contract simply because it had been priced as a one Aztec operation, but frequently a second aircraft had to be deployed to carry the weight of newspapers. It

Courtesy of Loganair

*Loganair operated the Glasgow–Stornoway newspaper contract for ten years from 1964. The Piper Aztec was used until 1968. The pilot routinely helped load and unload the newspapers. On the return leg from Stornoway, freight, often Harris Tweed, was carried on the aircraft. Here John Macdonald, 'John the Chemist', Loganair's Agent in Stornoway, is helping to unload.*

was minuted, for example, that in the month of November 1969, on fourteen occasions out of twenty-five, a second aircraft had to be deployed. The economics of the operation suffered accordingly. The Contract had become an albatross around the Company's neck. It was never a money-spinner.

The CAA Study Report of 1974 noted that losses on the Stornoway service as a whole accounted for virtually all the deficit in Loganair's charter operations in 1972/73. Loganair's truly remarkable and much publicised record of not having missed a single day's delivery in the ten years of the contract certainly established Captain Mac and his pilot colleagues in the folklore of Scottish aviation for their flying skills and service reliability, but for much of the ten-year period of the Newspaper Contract there had been plenty of kudos for Loganair but precious little financial reward. By 1974 the much heralded Stornoway newspaper operation clearly had run its course.

### The first 'scheduled service'

In October 1963, Loganair embarked on its first tentative escapade into 'scheduled' services which coincided with Duncan Logan (Contractors) Limited working on the Tay Road Bridge contract. Loganair had been instrumental in establishing an airstrip at Riverside Park, Dundee, and was flying Logan personnel involved in the bridge contract on a regular basis in the five-seat Aztec between Dundee and Turnhouse Airport in Edinburgh.

Duncan McIntosh believed that there was a demand from the wider Tayside business community for flights to Edinburgh and to join onward flights from Edinburgh to London and the south, and so a 'scheduled' service was offered. For thirty shillings one way, passengers could take the twelve to fifteen minute flight instead of a two and three quarter hour surface journey by road and two ferries, or the two and a half hour journey time by road or rail. This Loganair service was operated on an 'if and when required' basis and it is reported that the single daily service was flown only if sufficient seats were filled, or if one or two passengers were prepared to bear the cost of £7 10s for chartering the entire aircraft. Timings to or from Dundee were left more or less for the passengers to decide. What could be better?

Perhaps unsurprisingly this proved not to be a successful modus operandi. Despite Loganair's enterprise, it was still a very small air taxi company, not an airline with an appropriate structure. All bookings were handled by a local travel agent in Dundee. The flights were not adequately advertised or promoted, and when the Forth Road Bridge opened in 1964, passenger numbers fell away and Loganair's first 'scheduled' service experiment came to an end.

## First skirmish with BEA

The Stornoway newspaper operation continued as the Company's 'banker'. After having dropped off the newspapers in Stornoway, it was Loganair's practice on the return leg to Glasgow to carry in the Aztec all sorts of freight, but particularly Harris Tweed, in order to generate much needed additional revenue. This entailed the Loganair pilot unloading and loading his own aircraft; and completing the load and trim sheets and all relevant documentation. Unfazed by its recent experience of attempting a passenger service on the short Dundee–Edinburgh route, application was made in October 1965 to the Air Transport Licensing Board (ATLB) for a licence to fly passengers back to Glasgow via Benbecula by 10am in order to make flight connections to the south. The BEA service, after all, did not depart from Stornoway until 1510 and passengers wishing to fly onwards from Glasgow required to stay overnight for the following day's flights. Loganair's proposed early morning flight would allow either a full day in Glasgow or allow the passenger to make same-day onward connection. It all seemed very neat and sensible from both the Loganair and the passenger perspective.

BEA, which operated a daily Viscount service from Glasgow to Stornoway, lodged a successful objection with the ATLB using arguments which it and its successors, British Airways, were to make time and time again at Licence Hearings during the next twenty years: 'There was insufficient passenger demand; there would be a

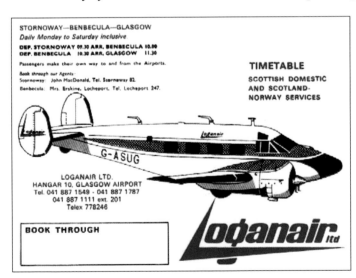

STORNOWAY—BENBECULA—GLASGOW
*Daily Monday to Saturday inclusive*
**DEP. STORNOWAY 09.30 ARR. BENBECULA 10.00**
**DEP. BENBECULA 10.30 ARR. GLASGOW 11.30**

Passengers make their own way to and from the Airports.

Book through our Agents:
Stornoway: John MacDonald, Tel. Stornoway 82.
Benbecula: Mrs. Erskine, Locheport. Tel. Locheport 247.

**TIMETABLE**

SCOTTISH DOMESTIC
AND SCOTLAND-
NORWAY SERVICES

G-ASUG

LOGANAIR LTD.
HANGAR 10, GLASGOW AIRPORT
Tel. 041 887 1549 - 041 887 1787
041 887 1111 ext. 201
Telex 778246

**BOOK THROUGH**

*Courtesy of Audrey Sanders*

*In 1966, the Company had won the licence to carry fifteen passengers per week on the Aztec's return leg from Stornoway after the newspaper flight. This often involved a request stop at Northton on its way from Stornoway to Benbecula. Here the advertised timetable in 1969 proudly depicts the Company's new Beech 18 aircraft.*

material diversion of passengers from the BEA scheduled service; these were primarily social services for local requirements; they were loss-making and likely to continue to be so; BEA had incurred losses of upwards of £5 million on Highlands and Islands services in the previous eighteen years and this vast investment would be jeopardised by Loganair's proposed five-seat Piper Aztec one-way service.' Needless to say, despite its obvious merit and the significant and demonstrable benefits that would accrue to the travelling public, Loganair's licence application was refused.

To their credit, the Loganair management persevered with the application. With the support of several local witnesses, including Alexander (Sandy) Matheson, later Convener of Western Isles Islands Council, and Lord Lieutenant of the County, who agreed to appear at the Licence Hearing before the Air Transport Licensing Board in London, Loganair's licence application was successful at the second attempt. Loganair's passenger service from Stornoway to Benbecula and onwards to Glasgow started in 1966 and continued for the duration of the newspaper contract until 1974, despite the licence restricting the Company to carrying a maximum of fifteen passengers per week.

Having at last won the licence to carry fare-paying passengers, Loganair was anxious to 'sell' its seats and services rather better than it had done with the earlier Dundee–Edinburgh service. The attractions of air services from Stornoway to Glasgow were self-evident. The traveller's alternative surface journey was arduous – by MacBrayne ferry, leaving Stornoway at midnight and sailing to Kyle and onwards to Mallaig. There, the passenger would take the train to Fort William and onwards to Glasgow on the West Highland Line, arriving at Glasgow at 4 o'clock the following afternoon.

In Stornoway, Loganair's important sales function was undertaken by 'John the Chemist' who was proprietor of Roderick Smith Limited, Pharmacy. The firm would buy all five seats on the flight and sell them to individual passengers. On Harris a similar function was carried on by Dr MacKinnon, and his passenger-customers would be uplifted from the beach at Northton with the Aztec landing there on a request stop on its way from Stornoway to Benbecula before flying onwards to Glasgow.

Loganair's passenger services were clearly flexible and informal. The stories relating to Loganair's operation of this scheduled service are legion, and Sandy Matheson remembers one occasion at Stornoway Airport when six booked passengers turned up for the five-seater flight. After much discussion, Captain Mac reached a compromise with one passenger who agreed to sit on the floor for the duration of the flight to Glasgow. At the end of the flight, and just when Captain Mac was congratulating himself on his masterly handling of a tricky situation, one of the other passengers announced that he was an Inspector for the Board of Trade. Happily Loganair got off with a severe warning.

Life, and fortunately also aviation, was more relaxed in these days as is also demonstrated in the arrangement to include Harris on the service. In 1967, Captain Mac received a telephone proposition from Angus MacKinnon, the doctor at Northton at the south-west tip of the Isle of Harris, who asked, 'Why not call in here?' Captain Mac commented that the large beach at Northton had been used before the war, but that was a long time ago. 'I'll give it a go,' he replied. 'I'll pop in tomorrow. Take your car on to the beach and make a few tyre marks and skids to let me see what the sand is like and, if it looks okay, I'll land.' This was the beginning of a regular arrangement, as Angus MacKinnon explains:

> He and I started the Piper Aztec landing on Northton beach and I used to do the bookings for Loganair from the doctor's house at Northton. I used to mark the touch-down spot by doing a figure-of-eight with my car tyres and the strip was marked with pink lobster floats kept in position with a weight on each one. I erected a wind sock in the middle of the expanse of sand and the pole, embedded in a barrel sunk in the sand, was visible there for years. The beach was not troubled by the tide, but large pools of rain water did gather on the sand and the winds would shift these about. While the beach was firm for landing and take-off, the wind would rock the aircraft while it was stationary and the wheels would dig themselves into the sand. I solved this problem with two large metal plates which the aircraft could use as an apron. The Piper Aztec, Islander and Beech 18 all landed at Northton during my time there from 1967 until 1972. My reward was the daily paper, wrapped in a plastic bag chucked out of the cockpit on the days the plane didn't have to land, and the occasional free trip to Glasgow.

These were the days. Needless to say, the MacKinnon/Northton runway test procedure was never adopted in the CAA Manuals.

### Serving Oban and Mull

The Dundee–Edinburgh service had perhaps whetted management's appetite for scheduled services and the Stornoway–Benbecula–Glasgow service was meanwhile seen as a real prospect. Thus encouraged, the Company busied itself trying to identify openings for new scheduled services and Oban and Mull came into Duncan McIntosh's line of sight. Loganair applied to the Air Transport Licensing Board for a licence to operate air services between Renfrew, Oban and Mull, a ninety-mile flight which took the traveller some eight hours by boat and train.

Before any air service could start, there was first the small matter of having suitable airstrips available. Argyll County Council had first to acquire the existing aerodrome at North Connel from the Ministry of Public Buildings and Works. A brand new airstrip had to be built on Mull and, in 1966, the Royal Engineers, under the OPMAC scheme, built a grass strip at Glenforsa in double-quick time. The Sappers felled more than 1,000 trees and moved 50,000 tonnes of soil in just fifty-four days.

There was much excitement in anticipation of the inaugural flight which would be operated by Captain Ken Foster. He was upstaged by Captain McIntosh flying in earlier with press men from Glasgow, and by Captain Geoff Rosenbloom who brought BBC cameramen. The press reported enthusiastically that Captain Foster's first official landing was greeted by a loud cheer from the watching crowd, including Provost MacLeod of Tobermory and many Council guests and a piper from the Argyll & Sutherland Highlanders playing *Chi Mi Muile*. After a celebratory lunch in the Glenforsa Hotel, the three Loganair pilots gave what was described in the press as 'an impromptu joy flight' for many of the excited guests. Heady stuff. Much effort then for a new air service which would be operated only at weekends and only during the summer season and which was to operate initially only until 1968.

### Orkney and the Britten-Norman Islander

Air taxi work, the Company's early core business, was clearly not achieving the financial results that were needed. Despite the Dundee–Edinburgh experience and Oban–Mull service which was proving to be anything but a financial success, Loganair's management believed more and more that their salvation lay in scheduled service operations. As early as 1965, the Company's attention was drawn to Orkney and, importantly, in this it was now being encouraged by the Scottish Office. This prospect clearly had much greater substance. Loganair duly applied to the Air Transport Licensing Board for a licence to operate scheduled services from Kirkwall to the North Isles of Orkney: Westray, Papa Westray, Sanday, Stronsay, Eday and North Ronaldsay. For once, BEA lodged no objection.

Very fortuitously, a new aircraft type, costing £26,000, was being manufactured at Bembridge in the Isle of Wight and seemed very suitable for this task. Captain McIntosh had flown, and indeed demonstrated, the new Britten-Norman Islander aircraft at the Farnborough Air Show in September 1966. He enthused:

> It can land and take off from an unprepared landing strip, like a farm field, and can cruise at 155mph. Compared with other similar aircraft, it can be operated fairly economically and the saving in costs that we make can be passed on to the passengers.

Captain Mac's enthusiasm for the aircraft was matched by no less a person than Group Captain Douglas Bader, reporting in newspaper coverage on the Farnborough Air Show:

*The Islander has played an important and, in the earlier period, vital role for forty-five of the Company's fifty years. It was instrumental in the Company securing the Scottish Air Ambulance Service contract in 1973 and even today fulfils a scheduled service role in Orkney.*

> Perhaps the highlight of the show was the Britten-Norman Islander, a twin-engined feeder liner which carries a pilot and nine passengers and can use unprepared airstrips. The small and enthusiastic factory at Bembridge, Isle of Wight, has already been in the news. They have received a number of orders for this grand little aeroplane and good luck to them.

Quite an endorsement from the great man.

Captain Fresson's Highland Airways had operated inter-island air services in the 1930s from Kirkwall to several of the North Isles, but these had been curtailed by the outbreak of the Second World War. There was a need for these air services to be resurrected for the benefit of the Orkney north isles which were totally dependent upon the ferry services of the Orkney Islands Shipping Company (OISC).

Duncan McIntosh clearly saw the suitability of the Britten-Norman Islander for ambulance work in Orkney as well as for the scheduled services: 'We can remove all the seats, and carry three stretchers at the one time and, of course, can carry any nursing staff who need to make the journey.' The Loganair plan was that the Islander aircraft would be based at Kirkwall, and would operate five scheduled services a day to different islands while also making links with BEA flights from Kirkwall to the mainland.

This time, Duncan McIntosh's optimism was well founded. The Company's Islander services would be a great success. Loganair's Orkney inter-island air services were operated initially under the aegis of the Orkney Islands Shipping Company which received the subsidy from the Scottish Office. It was the OISC which determined air service frequency and fare levels, and this arrangement prevailed until 1977 when Loganair assumed operation in its own right. Such was the popularity of the air services that no subsidy was required after 1971. During the initial period, Loganair's Islander carried the OISC funnel-flash on the tailfin, a red band with a white circle containing the letter 'I' signifying *Orkney Islands*.

## *Air Ambulance*

The service had started in 1933 with Captain Jimmy Orrell and the de Havilland DH84 Dragons of Midland and Scottish Air Ferries, fitted only with the simplest of avionics and certainly no de-icing equipment. Further north, Captain Fresson of Highland Airways was soon pioneering air ambulance flights in the Orkney Islands, followed in Shetland by Captain Henry Vallance of Allied Airways. In these pre-National Health Service days, patients had to pay for the charter of the ambulance aircraft, although various schemes were soon set up to assist with the significant cost. The pilots of these days were true pioneers and, deservedly, are still revered figures of the aviation folklore in the Scottish islands – Jimmy Orrell, John Rae, John Hankins, Eric Starling, David Barclay and Ted Fresson. With the arrival of the

Photo: Captain Eric Starling

Courtesy of Captain Duncan McIntosh

*The award of a supplementary air ambulance contract to Loganair in 1967 enabled the Company to expand the service which BEA's de Havilland Heron offered to islands such as Barra. The Britten-Norman Islander was to be key to Loganair ambulance operations, but the first air ambulance flight was operated to Oronsay with a Piper Aztec.*

NHS in 1948, the Scottish Air Ambulance Service became fully funded by the Department of Health for Scotland – later to be the Scottish Home and Health Department. In 1946, the independent airlines were nationalised and this took full effect in Scotland in 1947. For the next twenty-five years, British European Airways took responsibility for the air ambulance service, initially using de Havilland Rapides and the occasional Dakota (DC3), and from 1955 de Havilland Herons which by the 1960s were providing the air ambulance cover using seven airfields in the Scottish Highlands and Islands. With the arrival in 1967 of Loganair's first Britten-Norman Islander, and its excellent short field performance, it was soon apparent that air ambulance cover could be extended to many more remote areas throughout Scotland.

Loganair's first air ambulance flight had in fact been with the Piper Aztec on 16 June 1967 when Captain Ken Foster flew a patient from Oronsay to Glasgow. The Aztec then regularly sub-contracted by BEA to allow air ambulance cover to be extended to Coll, Colonsay, Oronsay, Mull and Oban. As Duncan McIntosh had recognised when it was first demonstrated at the Farnborough Air Show, the new Britten-Norman Islander could operate safely into some of the most basic airstrips. For the next six years, Loganair was able to play a supporting and supplementary role to BEA. This arrangement operated until 1973 when the Company was awarded the full contract by the Scottish Home and Health Department upon BEA's withdrawal of its Herons from service. Such was the versatility of the Britten-Norman Islander, to say nothing of the skill and courage of the Loganair pilots, that, in the first two years of air ambulance operations, some fifty airstrips were used.

### The team
Most importantly, Duncan McIntosh had gathered around him a team of pilots, almost all ex-RAF or Fleet Air Arm, who shared his enthusiasm and commitment, many of whom would continue with the Company and grow in stature throughout the next twenty years. As early as May 1963, Captain Ken Foster had joined the Company from the RAF. He would become Chief Pilot and later Flight Operations Director of the Company. There were other pilots too from these early years who became household names in different parts of the Loganair network. Captain Jim Lee, and later Captain Andy Alsop, both did much to establish and develop the Orkney inter-island air service. Captain Maurice (Barney) Barron, helped Captain Alan Whitfield establish air services in Shetland and was later a stalwart of the Inverness–Edinburgh scheduled service. Captain Geoffrey Rosenbloom and Captain Bill Henley shared the flying tasks later in the 1960s with Duncan McIntosh and Ken Foster on newspaper and ambulance work.

*Captain Mac's team in 1968. From left to right: Geoff Rosenbloom, Maurice (Barney) Barron, Ken Foster, Duncan McIntosh and Bill Henley – all were ex-RAF or ex-Fleet Air Arm pilots and real pioneers prepared to take on any flying task that presented itself.*

In 1963, a young local motor mechanic in Dingwall, Gilbert Fraser, caught Willie Logan's eye and was invited south by him to join the air taxi operation at Renfrew Airport. Gilbert progressed from the hangar floor to become Engineering Director in 1975. All of these men played an important role in the early pioneering days of Loganair. They flew aircraft to many of the remote communities of Scotland for the first time using new, very basic airstrips for charters and even more frequently on ambulance missions. Many families had good reason to be thankful to Loganair. Duncan McIntosh, Ken Foster and their pilot colleagues were real pioneers – brave, but always professional. They provided a lifeline to many remote communities throughout the Scottish Highlands and Islands.

### Progress to date

The Logan years had seen the Company take its first faltering steps in aviation. Its original *raison d'être* was to fly Willie Logan and the Duncan Logan management and staff between their many construction sites around the country, enabling them to make more efficient use of their time. As far as Willie Logan was concerned, 'time was money' and the air taxi company had served his own purposes very well. The construction company's flying requirements also provided the platform for Loganair to expand, as indeed it had to, and to make some progress even if much of this was by trial and error.

As a local Scottish aviation company, during this period Loganair attracted much positive publicity, firstly due to the high profile activities and business success of Willie Logan himself, and later by the feats of 'derring do' of Duncan McIntosh and his pilot colleagues. This was especially true on the west coast, with ambulance flying and the Stornoway newspaper contract. Loganair captured the public's imagination and generated a great deal of goodwill. Unfortunately, goodwill by itself does not pay the rent, and profitability proved very elusive throughout the period.

The Loganair management learned very quickly that it needed to fly the Company's aircraft harder if they were to improve the Company's overall financial position. In a totally Government-regulated flying environment, access to the existing scheduled service routes, which were very much the domain of British European Airways, had simply not been possible. As a consequence, charter opportunities had been pursued vigorously and the wide variety of work undertaken showed management's willingness to undertake most tasks which came its way.

Duncan McIntosh's first attempts at a rudimentary scheduled service on the Dundee–Edinburgh and on the Glasgow–Oban–Mull routes were probably always likely to fail. Operating the aircraft was one thing. Attracting enough customers quickly enough to ensure viability was another matter. Without adequate advertising and promotion of the service this would not be achieved. Loganair learned the hard way that it is always difficult for any airline to forecast with any confidence the likely passenger demand in the early years of a completely new scheduled air service. In the 1960s, this was nigh impossible for a fledgling 'airline' like Loganair with no previous expertise in this type of operation. Captain McIntosh did acknowledge with commendable honesty that route assessments were more 'by trial and error', but his willingness and indeed his enthusiasm for 'experimenting' paved the way for real scheduled service development later.

The Company regarded the Stornoway newspaper contract as its 'blue ribbon' flying. Certainly it was very significant in establishing the Company's reputation for commitment and professionalism. At the end of 1965, worryingly, it had become 'too important to lose'. The fundamental problem for the Company was the ever-increasing weight of the newspapers which frequently required the Company to deploy a second aircraft with no additional return.

Throughout the period, Duncan McIntosh regularly expressed his disappointment, and indeed surprise, at the low demand for air services in Scotland. He saw clearly the enormous advantage that air travel had over surface alternatives in the Scotland of the 1960s, but failed to realise that public acceptance of air charter and air taxi services inevitably took time. Scotland had no recent air travel tradition. It was nearly thirty years since the halcyon days of Captain Fresson and Gandar Dower who had flown their services from Longman and Dyce airports, and currently only BEA was flying what were very limited services to the Western and Northern Isles. Orkney would be the exception but it would take some time for Scottish passengers in any numbers to espouse air travel and to change their traditional ferry, rail and bus travelling habits.

There were a limited number of charter opportunities on offer, and several small aviation companies chasing them. The Loganair management became perhaps overly concerned about competition from other operators like Strathair, with one Aztec, one Helio Courier and an Aztec on order, and Gregory Air Taxis, based in Newcastle, and which was already taking some work which Duncan McIntosh thought rightly should have been Loganair's. The prices for flying jobs, therefore, were pitched very keenly, perhaps too keenly.

The arrival of the Britten-Norman Islander aircraft in 1967 undoubtedly was the most auspicious event in the fortunes of the Company during the Logan years. The operational performance of the new aircraft quickly made it eminently suitable for air ambulance work and enabled Loganair to become more and more involved in ambulance flying, first playing a supplementary role to BEA from 1967 and later to become the natural successor to BEA in 1973. The ambulance flying during the early period clearly was financially worthwhile. Through the operational flexibility, the ruggedness of the aircraft, and the experience and professional expertise of the Loganair pilots, air ambulance flying had the potential to grow significantly in size. And so it would prove. Over the years the Scottish Air Ambulance Service grew from strength to strength.

The ground work for successful air services in Orkney was also in place by the end of the 'Logan' years. Following protracted dialogue with the Scottish Office over a two-year period, Loganair's Britten-Norman Islander was seen to be ideal for Orkney's North Isles air services. These scheduled services were clearly very different from Loganair's earlier experiments on the Dundee–Edinburgh and Glasgow–Oban–Mull routes. These Orkney services were addressing a real passenger transport need; were themselves properly structured and, importantly, properly remunerated with the Scottish Office, through the Orkney Islands Shipping Company, offering contractual safeguards against any financial losses incurred. This was a valuable pointer – a blueprint for Loganair's future development.

### Parting of the ways

The period from 1962 to 1968 was loss-making for Loganair. However, by 1968 the Company had come a long way from the tiny air taxi company of 1962. It was now a highly respected 'third level' operator of a whole range of air services in Scotland. With increasing ambulance and Orkney operations, and with apparently encouraging dialogue beginning to take place with BEA, there were real reasons for optimism for the future.

With Willie Logan gone, however, Duncan Logan (Contractors) Limited was now considering its own future strategy. Serious cash flow problems were being experienced with its Kingston Bridge contract. Piling at the site had caused delays and cost increases due to the unexpected depth of glacial deposits under the River Clyde which had to be penetrated before solid bedrock was reached.

Clearly after much heart-searching, Martin Hill, Duncan Logan's Managing Director, wrote to Captain McIntosh on 11 October 1968:

> We, in Duncan Logan, have taken great pleasure and pride in assisting the development of Loganair from its infancy to its present position as Scotland's leading light aircraft charter company. Of course, this development would not have been possible were it not for the skill and enthusiasm of the team so ably led by yourself.
>
> Now you have reached a stage where we clearly see that there exists great potential for the further development of Loganair, and we have had to consider whether our backing should be applied to such development or to the expansion of our traditional business of construction. We decided to concentrate on the latter, provided we could find someone with the necessary financial resources, enthusiasm and vision to support Loganair in its progress in the market which we believe exists for the services you provide. We – and you – can count ourselves fortunate that the National Commercial Bank have come forward to undertake this task.

It was Willie Logan, and after his death in 1966, the support of the management of the Holding Company, that had made it possible for Loganair to survive as a company and to progress. Sadly, it was Duncan Logan (Contractors) Limited which would succumb first, never recovering from its difficulties with the Kingston Bridge contract. It went into liquidation in 1975.

For Loganair, and for Captain McIntosh and his colleagues, the change of ownership to an institution like the National Commercial Bank was very exciting. It clearly offered a more financially stable future with real scope for future expansion.

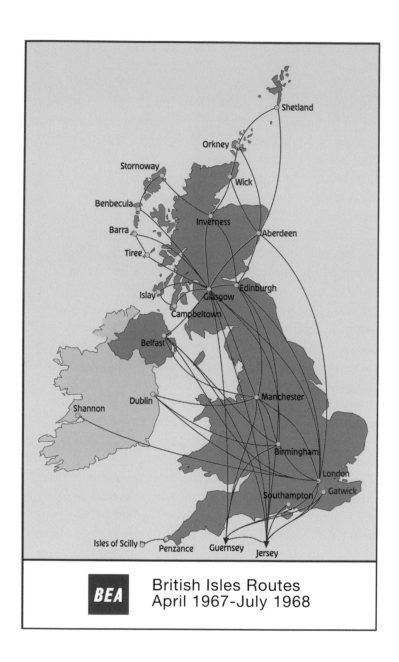

**British Isles Routes**
**April 1967–July 1968**

*Scheduled air services were regulated by the Air Transport Licensing Board until 1972 when this role was taken over by the Civil Aviation Authority. British European Airways had been operating, and incurring serious financial losses, on its Scottish routes since 1947. Loganair, with its smaller aircraft had aspired to take over some of BEA's "third level" services. Various Government Reports, notably the Edwards Committee Report, advocated change in air services in Scotland. Owned by the Royal Bank of Scotland, Loganair would surely be well placed to benefit.*

# Chapter 2

# The 'Bank' Years
# 1968 – 1983

## *Takeover by the National Commercial Bank*

On Tuesday 8 October 1968, the Share Capital of Loganair Limited was acquired by the National Commercial Bank of Scotland. Mr J C (Hamish) Robertson became Chairman of Loganair Limited with Captain Duncan McIntosh continuing as Managing Director.

Much was made of the benefit and convenience for the Bank of having aircraft to move its personnel between its 400 branches around the country. It was already in the public domain that the Bank would be merging the following year with the Royal Bank of Scotland and the number of branches in the Group would increase to around 700. Nor was the Bank slow to point out that it had been innovators in the banking field in the past and that owning a small Scottish Airline was quite compatible with its core banking activities.

The Bank claimed several world 'firsts'. In 1949, the Commercial Bank of Scotland had been the first to establish a Mobile Bank, on the Isle of Lewis. Following its merger with the National Bank of Scotland in 1959 to form the National Commercial Bank of Scotland, it introduced a Boat Bank in 1962, 'The Otter Bank', in the North Isles of Orkney, and it opened the first Ladies Bank in Edinburgh's Princes Street in 1964. The Boat Bank was to be superseded by the air service in Orkney in 1969 and became, along with its staff, something of an institution. Taking over Loganair was presented as a natural progression. No doubt the airline could have a useful role to play in the Bank's internal communications, but there is evidence that the Bank saw wider benefits in owning Loganair.

By 1968, Loganair was a well-known name and highly respected, especially in the Western Isles. It was emerging as an essential transport provider in Orkney and it was assumed already that internal air services in Shetland would soon follow the Orkney model. The Company was now playing a greater, albeit still supporting, role to BEA in air ambulance operations and the Bank had noted that all of these developments were with the encouragement of the Scottish Office. There had been the active role played by the Scottish Office in using the Orkney Islands Shipping Company as a mechanism to enable Loganair to become established in Orkney. The Scottish Office hopefully would now also be helpful in discussions which were now taking place with

BEA about the future shape of Scotland's air services. Loganair's routes and services were mostly 'lifeline' and socially popular. The Bank could thus identify a number of positive features in the Company it was acquiring.

The Bank certainly had ambitions for its new airline. In the same week as the takeover, Ian MacDonald, Chairman of the National Commercial Bank, stated: 'Our ideas may be that Loganair may well provide feeder services for BEA and BOAC and take care of operations into small landing strips elsewhere. With our operations we plan to open up Scotland.' It was evangelical and ambitious. It might even have been logical as it was known that BEA's Herons and Viscounts would eventually have to be replaced and there were no suitable replacement aircraft in view. Ian MacDonald and the Bank, however, had not reckoned on the mindset of the BEA management, but at least in the meantime there would be kudos for the Bank in being associated with these regional, 'lifeline' air services. Profitability would surely follow.

### Chairman Mr J C Robertson

As one would expect, Chairman Hamish Robertson started in a brisk, business-like fashion. He decreed there would be monthly Board Meetings for which monthly Management Accounts would be prepared for the first time. Mr Jim M Harter, of auditors David Strathie, had been appointed Director and Financial Advisor. Later Mr Robert B Smith was seconded as Company Secretary from the Bank. Hamish Robertson immediately needed to determine the most appropriate capitalisation for the Company. This would clearly depend on the funding requirements for the Company's future expansion which, of course, would now be the subject of policy decisions to be made by the Bank. The Authorised Share Capital was increased to £500,000 with £350,000 Issued, as the total funding requirements were deemed to be £350,000 in respect of working capital and, significantly, £170,000 for the purchase of a new Short Skyvan aircraft.

At his first Board Meeting, Chairman Hamish Robertson set out his stall. It was his intention to approach Lloyds & Scottish Finance Limited for favourable terms for the Company's early repayment of its various Hire Purchase Agreements – henceforth all funding requirements would be met by way of bank overdraft or loan facilities. He instructed that the Company's order for its third Britten-Norman Islander be cancelled. Any surplus funds would be deposited at short notice with National Commercial & Glyn's or Schroders Limited. He would approach the Bank to explore Group Relief, whereby, as a wholly owned subsidiary, Loganair's tax losses and capital allowances which had arisen through its trading losses and aircraft acquisitions, could be surrendered to the Bank to be used to reduce the Bank's own Corporation Tax liability. The Bank in turn would pay Loganair the cash equivalent of the tax saved. This practice was to be repeated throughout the period of the Bank's ownership and would be critical to the Company's cash flow. The first Group Relief payment of the significant sum of £169,000 was made by the Bank to Loganair in October 1970. All this made sound financial sense for Loganair. Hamish Robertson and Jim Harter were determined to put the Company on to a sound financial footing which would make expansion a real possibility.

### Aircraft Fleet Development

Historically, prior to buying aircraft, assessments were carried out by the Company, but these were technical, operational performance assessments carried out primarily by Captain McIntosh and Captain Ken Foster, the Chief Pilot, to determine the aircraft's suitability for the flying task in hand. Much less precise in the assessment process was the aircraft's forecast annual utilisation and, even less precise, the prospects of a financial return from the aircraft's use. This was not the Bank's usual way of doing business, but Chairman Hamish Robertson would learn the hard way. Within a week of the Commercial Bank taking over Loganair, a Beech 18 was delivered, having been ordered some months earlier when Chairman Martin Hill had bowed to pressure from management as well as the newspaper distributors.

The Beech 18 was perhaps more popular with the pilots than it was with the passengers. 'This is a real pilot's aeroplane, and thoroughly rewarding to handle,' Captain McIntosh told the press. He always called it, with genuine affection and admiration, 'The Queen of the Skies'. The Company's management never lost faith

*The Beech 18, G-ASUG, was acquired in 1968 to carry passengers and also to carry the increasing weight of newspapers on the Glasgow–Stornoway contract. It was also used on the Company's first international scheduled service on the Aberdeen–Bergen route in 1969/70.*

Courtesy of Loganair

in it even although the aircraft was often unserviceable for lengthy periods. In October 1972, when undertaking a mandatory main spar modification, Loganair decided at the same time to carry out an expensive interior refurbishment and to convert it into a seven-seat executive passenger aircraft – but this change in configuration brought no change in fortune. In 1975, after strenuous efforts to find a purchaser, and finally having sold engines and other components, Beech 18 G-ASUG was gifted by Loganair to the Museum of Flight, East Fortune, at the ripe old age of twenty-two years, and where it proudly remains today alongside a British Airways Concorde.

Hamish Robertson had inherited the Beech 18, but worse was to follow. At his very first Board Meeting on 17 December 1968, almost before he had his feet under the desk, he was caught cold. Captain McIntosh was reporting that charters were having to be refused due to shortage of aircraft and he felt that 'notwithstanding the introduction into service of the Beech aircraft, additional work might be available if the Company had an additional aircraft and the Short Skyvan might be considered.' An 'assessment' was ordered to be carried out, and very soon thereafter, on 3 March 1969, Short SC7 Skyvan, G-AWYG, was delivered – the first to a UK operator.

The Skyvan, with a payload of up to eighteen passengers or 4,000lbs of cargo, was

Courtesy of Loganair

*The delivery of the Shorts Skyvan on 3 March 1969 was an exciting milestone in the Company's development. New owners, Commercial Bank of Scotland, naturally took a keen interest. Left to right in the photograph: Ian MacDonald, Chairman of the Bank; Jim Harter, Loganair's Director and Financial Advisor; Denis Tayler, Manager of Shorts Skyvan Division; Hamish Robertson, Chairman of Loganair; Captain Duncan McIntosh, Loganair's Managing Director; Alex Roberts, Short Brothers' Sales Director and John B Burke, General Manager of the Royal Bank of Scotland.*

Loganair's largest aircraft to date. Its STOL capability, the Company management were convinced, would make it 'eminently suitable' for the smaller Highlands and Islands airstrips. The argument ran that the Skyvan would bring much new air freight business. Its squat box shape allowed it to carry both bulky and heavy loads and its rear door meant it was in a class of its own in its ability to carry unorthodox items. There were indeed a number of such loads which the Company ensured were given the full press treatment. There was the Beach Buggy vehicle, which just happened to belong to Loganair pilot Geoff Rosenbloom, carried to an Autocross event in Inverness. Better still, Loganair was able to transport a Land Rover to the Island of Coll. Unfortunately there was not much demand for transporting Land Rovers. Again amid much publicity, the Skyvan, with its convenient rear door, had been chartered by the Norwegian Parachute Regiment whose activities included a stunt involving parachutists on bicycles making a jump from the aircraft at 8,000 feet. The Skyvan was to prove its real worth, however, when it carried from Glasgow to Staverton in Gloucestershire the forging for the nose gear of the prototype Lockheed Tristar which had been cast at the Carron Ironworks in Falkirk.

All too soon, however, the Company discovered that its old problem of low aircraft utilisation was particularly acute with the Skyvan. Costings had been based on an expected annual utilisation of 800 hours and, after a year, actual utilisation was rather less than half of that target – and this despite the Skyvan sometimes flying very frequently on the Stornoway newspaper contract when newspaper weights exceeded the carrying capacity of 2,200lbs for the Beech 18. The Skyvan was also used on occasion as back-up for the Beech 18 on the Glasgow–Aberdeen–Stavanger service, launched in 1969. However, new charter work for the Skyvan was hard to come by and very soon this increasingly became a source of real concern to the management. The aircraft was expensive to operate, and to make matters worse, on longer journeys its payload could be restricted to as few as twelve passengers through the need to carry round-trip fuel because the turboprop Skyvan used kerosene fuel which was available only at the major Scottish airports.

As early as February 1970, less than a year after delivery, serious consideration was being given to mothballing the aircraft in an attempt to reduce the Company's financial losses. However, it would be five years before it left the Company's fleet and be sold to the Omni Aircraft Corporation in Baltimore, USA. Short Skyvan 3 G-AWYG had been an expensive mistake for Loganair and would continue to be a financial burden on the Company right up to the time it was sold. No one would confirm the report that the final decision to part with the aircraft was taken after a Skyvan charter carried homing pigeons from Glasgow to the Faroe Islands, and all the birds arrived back in Glasgow well ahead of the aircraft.

Courtesy of GAAEC

*The Shorts Skyvan followed the Beech 18 into the Company's fleet. Its increased passenger seating and its suitability for freight and especially heavy lifts was never fully exploited notwithstanding the photograph of the reluctant bull. The Shorts Skyvan proved to be a serious financial drain on the Company.*

Courtesy of
Captain Ken Foster

Photo: Stuart Sim

### An International Scheduled Service!

The early years of the Bank's ownership saw the Company's financial losses continue. The expansion of the business had been too rapid. Two larger aircraft, the Beech 18 and the Short Skyvan, had been added suddenly to the fleet of two Aztecs and two Britten-Norman Islanders, a significant increase in capacity in a very short period. Much new, remunerative work was needed quickly. Nonetheless, applying for a scheduled service licence for the Glasgow–Aberdeen–Stavanger route was most ambitious. The Company had been encouraged to start this service by James Davidson, the Liberal MP for West Aberdeenshire. The licence was granted on 19 June 1969 with the service commencing much too soon thereafter, in early July. Captains Ken Foster and Geoffrey Rosenbloom flew the Skyvan on the inaugural flight with Chairman Hamish Robertson on board, together with a piper from the Bon Accord Ladies Pipe Band to play *Scotland the Brave* at Stavanger Airport.

The inaugural flight, at least, was a great success, and attracted much publicity and indeed some controversy. The Company had applied to the International Air Transport Association (IATA) to charge a single fare of £25 on the route, but BEA had immediately lodged an objection. As a result, Loganair had to charge £29 10s each way as that was the current fare of carriers linking these points, albeit over London. David Steel MP actually tabled a question in the House of Commons and commented: 'This is a first rate piece of Scottish enterprise which could come to nothing because it is strangled in red tape. Why should the routes all go through London anyway?' Why indeed?

The service started with high hopes and initially passenger revenue just covered the direct operating costs of the route. The Company had sensibly timed its flights to arrive at Stavanger to allow connections to be made with internal Norwegian flights to Oslo and Trondheim. It all seemed very positive, but even on the inaugural flight there were Norwegian complaints that the fares were too high. Within three months passenger carryings were declining and the winter service was reduced to two days per week, Mondays and Fridays. The Company had decided not to engage Rex Stewart and Associates (Scotland) Limited, nor to accept their proposals for publicity and advertising, as their quote of £435 was considered to be excessive. This may have well been a serious error of judgment.

By February 1970, Chairman Hamish Robertson was expressing serious concern about the service and intimated that unless there was an improvement the service would be cancelled. With the services restricted to twice per week, results did improve to the point that on several days in July and August, the fourteen-seat Skyvan was used because the Beech 18, limited to a seven-seat capacity, could not carry all the passengers. It was hardly a new dawn, however, and by October passenger traffic and

*Aberdeen-Stavanger Inaugural Flight – 14 July 1969. Chairman Hamish Robertson, sixth from right, accompanied by BBC and STV crews and reporters and by the piper from the Bon Accord Ladies Pipe Band. The party was met at Stavanger by the UK Consul and local officials. The service was the first new international service from Aberdeen Airport for more than thirty years. Unfortunately, North Sea oil exploration did not start for another three years and meanwhile the Loganair service had had to be discontinued in November 1970 due to lack of passenger support.*

forward bookings had again deteriorated. In November 1970, with only six bookings taken from then until the end of the year, the Board decided that enough was enough and that Loganair's first international service should be discontinued.

The Glasgow–Aberdeen–Stavanger scheduled service had been a big step for Loganair, demonstrating management's commendable ambition and a characteristic 'can-do' attitude. It had hoped that the service would be used by tourists from both ends of the route, as well as by Glasgow and Aberdeen businessmen. During the sixteen months of operation, the Board had been particularly surprised that the Stavanger route had generated little traffic from Scotland and some seventy-five percent of the custom had come from Scandinavia.

It had been a step too far, however, as the Company was not yet geared to make a commercial success of a scheduled service in Scotland let alone an international passenger service and passenger demand had been modest throughout. The Company really was ahead of its time – the new service to Stavanger was the first new international scheduled route to be operated from Aberdeen by any airline for thirty years, and sadly it was before the market was quite ready. Ironically, it was only three years before the oil boom in the North Sea took off, and the fortunes of many airlines dramatically improved. But not for Loganair. 'We lost a bomb,' Captain McIntosh told the press when announcing the end of the service.

### *Back to Dundee*

If a scheduled service to Stavanger was new and exciting for the Company, starting services again to Oban and Mull, and also to Dundee were decidedly déjà vu. Since Loganair had put Dundee Riverside on the aviation map in 1963 with its brief experiment on the Dundee–Edinburgh route, Dundee had seen other operators. British Eagle had flown from Riverside Park to Renfrew using a de Havilland Dove, and Autair International operated from nearby Leuchars to London with a Hawker Siddeley 748, but these were all short-lived. By 1970 the city of Dundee was again without an air service.

The Provost of Dundee personally approached Duncan McIntosh and requested a twice-daily feeder service each weekday on the Dundee–Glasgow route as Glasgow had a wide range of onward services providing greater potential for interlining traffic. Riverside airstrip had been closed in 1968 after the Corporation had decided to sell ground to the University of Dundee as a playing field in order to offset the cost of the £15,000 subsidy paid to Autair, but at an emergency meeting the Council decided to retain the grounds as an airstrip.

Early discussions about a new service were confused by the possibility of British United Airways guaranteeing four seats on each flight to connect with their Glasgow–London Gatwick scheduled services. These discussions came to nothing and Loganair requested a subsidy of £30,000 from Dundee Corporation. After protracted negotiations, the Britten-Norman Islander service started on 26 June 1970 using Leuchars because Riverside Airport was unserviceable due to water-logging. Initial passenger demand was disappointing and, even with subsidy, continuing the service beyond six months was problematical. It was only when the service was able to switch to Riverside, conveniently situated only a mile from the city centre, that passenger carryings improved.

The Dundee service was a struggle for the Company. The Corporation agreed to a subsidy of £20,000 and, by the end of 1972, passenger demand was encouraging. The Corporation even requested Loganair to look at extending the service from Dundee to Aberdeen. Needless to say, the Corporation turned down the Company's request for additional subsidy. The Company was trying to be innovative. It introduced a third, mid-morning service on the Dundee–Glasgow route without subsidy in view of the reported number of passengers having to be waitlisted due to full flights. Later, the sixteen-seat Trislander aircraft was introduced on the route and Loganair again unsuccessfully sought extra subsidy. However, passenger traffic drifted downwards and was further affected by British Caledonian's decision in March 1974 to provide a coach service from Dundee to Edinburgh to connect with its London service. Then interlining traffic suffered when British Airways changed its Glasgow–Birmingham timetable.

The Company was still negotiating for subsidy when Riverside Airport again became flooded and rutted. The Loganair Board, confronted by an apparent indifference from Tayside Regional Council who had just taken over responsibility for the airport from Dundee Corporation, terminated the service on 15 May 1975 after five years of considerable effort, endless negotiation over subsidy, and considerable frustration for Loganair.

### Oban–Mull Again

The Company fared no better when it revisited another of its old routes. Loganair had been operating from 1 September 1967 on its former route to Glenforsa in Mull on behalf of the small charter company Benmore Flights Limited, owned by Viscount Massareene. When Benmore Flights had no plans to operate in 1969, Loganair applied for and was awarded a licence to operate three days per week during the summer months. When Loganair declined Viscount Massareene's invitation to take Benmore Flights Ltd over, that company went into liquidation. Loganair's own service started in June 1969, but passenger carryings were to be very disappointing throughout the season. A reduced frequency was operated the following summer and again in 1971 with weekend services in peak summer months only.

Even when it was clear that the Highlands and Islands Development Board (HIDB) would not be permitted to subsidise the route, the Company rather surprisingly continued to operate the service the following summer, still with no prospects of passenger growth despite the service being extended from Glenforsa to the island of Coll. These services were operated in the relaxed, informal manner of the day. The fireman at Glenforsa would park his red tractor at a specific location as a signal to the pilot on his return flight from Coll that there were passengers to be picked up for Glasgow. Charles Randak remembers that, on one memorable occasion at least, the pilot duly landed back at Glenforsa to be told by the fireman that, in fact, there were no passengers waiting, but would he take this freshly caught salmon back for Gilbert in the Glasgow hangar.

The service was marginal at best and often failed to cover the aircraft's direct operating costs. Eventually, the Board admitted that the Oban-Mull flights were being operated merely to keep a Loganair presence on the West Coast during the summer season of 1974.

One could be forgiven for thinking that Loganair's efforts to establish a viable service to Oban and Mull through until 1975 was above and beyond the call of duty. Once bitten, twice shy. Not a bit of it. Captain McIntosh had one last attempt in 1978 at operating a daily service, Monday to Saturday, but this last ditch attempt at a higher frequency proved no more successful than the previous Mull services and lasted only two seasons. Later, in a warm personal tribute to Captain Mac, his friend and colleague Captain Alan Whitfield wrote, 'He persisted with the Mull service for a long time because I think he had a very soft spot for it.' That's just the way it was.

### Hamish Robertson's Stewardship

Hamish Robertson stood down as Chairman of Loganair on 31 March 1972 upon his retirement from the Bank. He had presided over the Company for three and a half years of serious financial losses. His task had been made extremely difficult by the sudden fleet expansion with the Company taking delivery of the Beech 18, quickly followed by the expensive new Short Skyvan, as soon as he became Chairman.

The Company never got close to the necessary annual utilisation or revenue needed to sustain them and inevitably serious financial losses ensued. Month after month, the Board had agonised about the under-used fleet, but now there were other seriously negative contributory factors at work. Not the least of these was the often hopelessly inadequate income from the Stornoway newspaper contract and the serious revenue shortfall on the Stavanger scheduled service. The political climate for change in Scottish aviation may have improved for Loganair during his time as Chairman, but for Hamish Robertson, effecting that change and turning the Company around had proved to be frustratingly slow.

### Government Reports and challenge to the status quo

Throughout these early years, Loganair had been endeavouring to identify and develop a viable and sustainable role in air transport in Scotland within a regulated industry which invariably protected British European Airways. To its credit, BEA had provided essential air services in the Highlands and Islands for social rather than commercial reasons since February 1947 that were loss-making and viewed at the time as likely always to be so. Nonetheless, it was not only Loganair which was now questioning the BEA-dominated regime. There was now a growing desire to explore how air services in Scotland might be improved. Recognising this, the Government sponsored a succession of important studies and these would lead to much needed change in Scotland's air transport services and infrastructure. Loganair would have to be patient, but these surveys would eventually create opportunities for Loganair to play a bigger role.

The first of these more than set the scene. The Highland Transport Board, the forerunner to the Highlands and Islands Development Board and later Highlands and Islands Enterprise, was appointed by the Secretary of State for Scotland in 1967 'to advise on the most economic ways of meeting the needs of the Highlands for transport services whether by land, by sea, or by air', and it published a Report, *Highland Transport Services*. The Report was critical of BEA's one round trip frequency on most air services north of Glasgow and Edinburgh, which manifestly failed to meet the requirements of regular travellers. At long last the status quo was being challenged and indeed a programme of positive actions quickly followed.

Although not helpful to Loganair's cause, the Highland Transport Board Report supported BEA's proposals to introduce Viscounts on the north-eastern and Glasgow-Campbeltown–Islay routes, currently served by Heralds, which they deemed not large enough to meet increasing traffic demand and which, significantly, were also due for major overhaul. Extensions to runways at Sumburgh, Kirkwall and Islay airports were required for the Viscount services and these were completed by the Board of Trade in 1966. BEA subsequently operated Viscounts on all routes in the Highlands and Islands except for services to Tiree and Barra for which the Heron was used. For a short while following withdrawal of the Herons, BEA also used Short Skyliners.

Of immense significance for Loganair, however, it was the Highland Transport Board which had recommended the resumption of the pre-war North Isles of Orkney services within the operating structure of the Orkney Islands Shipping Company which then entered into a charter arrangement with Loganair. Improvements to existing airstrips and the construction of several new airstrips in the North Isles were paid for by Orkney County Council with a grant from the Board of Trade. Loganair commenced these services in September 1967. Finally the Highland Transport Board recommended that Inverness and Wick aerodromes be staffed to enable early morning and late evening services to be operated, consideration at last for the day return services which passengers actually wanted.

### The Edwards Committee Report

Progress indeed. It was the next Report, however, the Edwards Committee Report in 1969 entitled *British Air Transport in the 1970s* that made the whole industry sit up and take notice. Loganair certainly saw it as its big chance. The Loganair Board pointed to many auspicious elements of the Edwards Report and talked up their relevance for Scottish air services and, of course, Loganair's eligibility. Loganair was one of a handful of small airlines referred to in the report: 'We believe that third level services of this kind will become increasingly important in this country during the next decade. It is a field in which small scale is a positive advantage, and these are activities which should be left for private airlines to develop.'

The Report also drew the distinction between 'Second Level' and 'Third Level' air services. Third level was generally understood to refer to local air services using aircraft with fewer than twenty seats. In the Scottish context, BEA clearly was the Second Level carrier and Loganair was the Third Level carrier. The Report also made clear that it disapproved of the hidden cross-subsidisation of unprofitable by profitable routes. The Edwards Committee argued that direct subsidy was appropriate if it was considered essential to maintain socially important, but commercially unprofitable, air services.

The Board, now with the might of the Royal Bank of Scotland behind it, very quickly told the Government, through the Scottish Development Department, that the Company, either alone or in partnership with the Highlands and Islands Development Board, was interested in taking over the services which had been costing BEA money for years. That was certainly true. BEA had been serving Scotland faithfully since 1947 and had incurred significant losses throughout the period. Loganair's temerity in talking to the Government in such a manner, however, drew a sharp riposte from Sir Anthony Milward, BEA's Chairman: 'BEA has no intention in giving up any of the services in Scotland.' His recent tour of the Highlands airfields, Sir Anthony declared, had shown him who the people really wanted to serve them: 'They wanted BEA.' Sir Anthony also chose to play down the £423,000 loss on its Scottish operations in the airline's Annual Report for the year ended March 1969, a point which was picked up in an editorial in the *Scotsman* newspaper: 'BEA is no longer parading the Highland services with the mien of the willing martyr. Underplaying the unprofitability is a new tactic.'

For Loganair, the Edwards Committee Report was a source of great encouragement. It also caught the mood of the moment. The Government was busy deciding in the run-up to the General Election in eighteen months time whether to write 'aviation devolution' for Scotland into the Queen's Speech at the Opening of Parliament. Meanwhile, the HIDB was also discussing Scottish aviation with the Secretary of State for Scotland. BEA was on the defensive. But then, for different reasons, it had been on the defensive for years.

### British European Airways

There had been much speculation for a number of years about the future shape of the UK's nationalised airlines. BEA had itself been fighting something of a rearguard action against a merger with BOAC. As early as 1962, Lord Douglas, the BEA Chairman, had felt it necessary to publish a comprehensive denial: 'The interests of

BEA and BOAC were too diverse to make a merger sensible.' When the long-awaited Edwards Committee Report was published in 1969, it was eagerly received by the politicians, but viewed with considerable suspicion by BEA. Admittedly, the Report had not recommended the merger of BEA and BOAC. Indeed, it suggested that the two airlines should retain their own identities, but it did recommend that a new holding company, the National Air Holdings Board, should be set up to supervise the running of BOAC, BEA and British Air Services (BAS), and Chairman Sir Anthony Milward feared that this new Holding Board could be a suitable vehicle for the Government to implement merger. In the 1969/70 Report and Accounts, he wrote that, 'There is little merit in any proposal to change for change's sake.' The Edwards Committee Report had also recommended a strong independent airline be encouraged to become 'a second force' to the nationalised airlines, and British United Airways and Caledonian Airways were suggested as possible suitable candidates. With all these possible changes in the offing, the BEA Board clearly were not in the mood to permit any concessions being made in Scotland to Loganair or anyone else.

And so it proved. To the intense disappointment and frustration of the Loganair Board, there would be no early route transfer, no abandonment of Scotland by BEA, and only a greater determination than before to preserve the status quo. BEA announced a new initiative, a 'profit centre' regrouping scheme which created a Scottish Airways Division from April 1971. The new Division would be responsible for the running of the Scottish internal services, but, to help balance the books, it would also include services linking Aberdeen and Inverness with London, and, for good measure, also the Glasgow–Belfast route.

BEA's Chairman, proclaimed that no routes in Scotland were under consideration for closure. In a departure from recent policy, he also stated that, if a service was vital socially, and impossible to make pay, the Government would be asked for a subsidy. Cyril Herring was to be the BEA Director responsible for Scottish Airways Division and Robert McKean, the Scottish Director. All BEA aircraft, including the eight Viscounts allotted to the Scottish routes, would bear the words *Scottish Airways* along the fuselage. BEA would make much of the fact that the new scheme was in line with the Edwards Committee Report which had suggested eighteen months earlier that Scotland should have a regional airline. So that was that. Or was it?

Despite its initiative in setting up Scottish Airways, and no doubt to its disappointment, BEA continued to come under pressure from the Government to consider the rationalisation of third level air services in Scotland. BEA's Scottish services were still incurring heavy financial losses and, although there was a clear case for subsidy for the Highlands and Islands services on social grounds, it was apprehensive about the implications of accepting subsidy. Something had to happen and the Scottish Development Department decided that a Joint Working Party be formed. The first meeting took place in January 1972 with BEA and Loganair representatives present. The routes from Glasgow to Tiree and Barra, Glasgow to Campbeltown and Islay, and from Wick to Kirkwall and Sumburgh were specifically within the scope of the Committee's remit, and, most importantly for Loganair, the future of the Scottish Air Ambulance Service contract. To Loganair's chagrin, it quickly became clear that BEA was merely going through the motions. There was no will, and certainly no appetite, on BEA's part for change.

While BEA may have been agonising over the question of subsidies, it was absolutely resolute in defending its interests in Scotland. Earlier it had formally objected to Loganair's application for a licence to carry passengers on the return leg from Stornoway to Glasgow using the newspaper charter aircraft. Now it was to lodge an objection to Loganair's licence application to operate on the Inverness–Edinburgh route on the grounds it might harm its Viscount service from Glasgow to Inverness which flew via Turnhouse. BEA management was now even more resistant to change. Even when it became clear that BEA had no aircraft suitable for air ambulance work, it was reluctant to give up the contract to Loganair, being more prepared to sub-contract the work and, even as late as 1971, BEA was considering its own helicopters for the ambulance role.

Even more disappointingly, even after decisions had been taken to withdraw the Heron and the Skyliner from service, BEA was prepared to consider subcontracting aircraft for the Tiree and Barra services rather than allow Loganair to operate the routes on its own. There was a strong feeling within Loganair at the time that the Scottish BEA management, Patrick Gillibrand, Robert McKean, and later, Hugh Reid, saw the merit of a closer collaboration with Loganair in Scotland, and had favoured handing over the Tiree and Barra scheduled services and the air ambulance contract to Loganair, but that they were not being permitted to do so by their masters in London. To the intense frustration of the Loganair Board, no real progress at all was made and Mr Cyril Herring of British Airways' Board would give no clear sign that cooperation would take place soon between Loganair and BEA. These were dark days.

Thoroughly disenchanted by this impasse and the BEA attitude, especially in relation to the air ambulance, the Loganair Board resolved to get on with it and to tender immediately for the full Scottish Ambulance contract. This time BEA had no hand to play as its two de Havilland Herons were being withdrawn from service. Clearly Loganair had a strong case anyway as the Company had demonstrated its ability to provide, and indeed extend, air ambulance cover for the last six years as subcontractor to BEA.

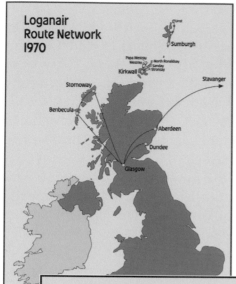

The Orkney Isles network is already well established, while Unst is operational as the first step in the Shetland Isles internal service. Passengers were carried from Stornoway and Benbecula on the return leg of the newspaper run and Dundee had a link with Glasgow operating from Leuchars. The most ambitious route was the international service linking Glasgow and Aberdeen with Stavanger.

Progress had been made since the Bank had taken over in 1968. The Orkney and Shetland inter-isles networks were already well established. In the Western Isles, passengers were being carried from Stornoway and Benbecula on the return leg of the newspaper run and Glasgow was still linked with Dundee. Highlands & Islands Development Board was financially supporting the services from Inverness to Aberdeen, Dornoch, Wick and Skye, which was also served from Glasgow. A summer service was now operating from Glasgow to Oban, Mull, Coll and Tiree. Loganair's early experiment with international services from Glasgow and Aberdeen to Stavanger had been short-lived and had been discontinued in 1970.

Even then, there were further twists. Because of the serious financial position that Loganair was then in, the Royal Bank of Scotland, as Loganair's owners, were asked to provide a guarantee of support for Loganair during the three-year contract. The Bank was only too pleased to make the commitment. Concern was then expressed that the Loganair management was not sufficiently strong, and it was agreed that a Financial Manager would be recruited. Finally, at the eleventh hour, on 26 January 1973, Gordon Campbell, Secretary of State for Scotland, announced that Loganair would be appointed sole operator of Scotland's Air Ambulance Service with effect from 1 April 1973. It was a red letter day for the Company and Duncan McIntosh was excited: 'We have arrived. And when we take over the air ambulance work, we will have the final accolade of respectability.' There may have been serious frustration felt towards the BEA hierarchy in London, but BEA Scottish Division Chairman, Robert McKean and his Scottish management colleagues were most gracious and offered Loganair every assistance.

Henry Marking may have set up Scottish Airways Division with a fanfare of trumpets, with its aircraft adorned with its new *Scottish Airways* branding, but it seems that it was not only Loganair which was sceptical. The Rt Hon Peter Walker, Secretary of State for Trade and Industry, in 1972 instructed Lord Boyd-Carpenter, Chairman of the Civil Aviation Authority to report on *Air Transport in the Scottish Highlands and Islands*. This was seen as hugely important and all interested parties, including, of course, Loganair, submitted evidence. To the acute frustration of the Loganair Board, the Report was not published until April 1974, and, in the interim, there was yet further confusion and frustration with BEA. Patrick Gillibrand, who headed up BEA in Scotland, intimated to John Burke that he was authorised to discuss acquiring a forty-nine percent shareholding in Loganair.

BEA then wanted to reconvene meetings in 1973, but when it became clear to the Loganair management that this sudden request for closer cooperation was probably designed to address BEA's under-staffing problems at some of its Highlands and Islands outstations, the Loganair Board remained unimpressed and decided meanwhile to stay its distance from BEA. Loganair would await the CAA report. It had become clear during the consultation that the CAA Study Group would be unhappy with BEA having an equity link with Loganair as it favoured Nationalised Concerns having effective competition.

### Meanwhile, an HIDB initiative

During the frustrating period following the Edwards Committee Report during which the BEA Board was being totally obdurate, the new Highlands and Islands Development Board was keen to promote new local air services throughout the

Highlands and Islands from a base at Dalcross Airport, Inverness. The Board offered Loganair a contract to provide services from Inverness to the new airstrip at Broadford on Skye, and from Skye to Glasgow; and additionally, services from Inverness to Aberdeen and north to Dornoch and Wick. Unfortunately passenger carryings on all the new services were modest, HIDB could not justify continuing its grant to Loganair beyond December 1973, and the services were discontinued except for the service between Skye and Glasgow.

HIDB's commendable effort had been planned to be the first stages of a wider air service network, where 'bus stop' services with small aircraft would link many communities around Caithness, Sutherland and Wester Ross. This grand plan was terminally damaged by the international oil crisis between October 1973 and March 1974, and the huge increase in the cost of aviation fuel.

The Skye–Glasgow service, however, continued until 1986 with subsidy from the Scottish Office. It was never well supported by the indigenous population of Skye and only had a 'purple patch' during the construction of Chevron's Ninian Platform at the Kishorn Yard in 1977/78 and when there was Naval Research work in connection with the testing of the Stingray torpedo in the Inner Sound of Raasay. When this activity ended, the passenger demand was too low to justify the subsidy being continued.

### Mr John B Burke

John Burke was educated at Hutcheson's Grammar School in Glasgow, served in the Navy during World War Two attached to the Fleet Air Arm and, on joining the Royal Bank of Scotland after the War his career progress had been meteoric. Even before he succeeded Hamish Robertson on 1 April 1972 as Chairman of Loganair, he had taken a keen interest in Loganair's fortunes. He himself was an aviation enthusiast and on becoming Chairman he took flying lessons to gain his Private Pilot's Licence. He liked and understood the business Loganair was in, but he was also unquestionably committed to getting the Bank's wholly owned subsidiary on to a profitable basis. He had been present in November 1969 when Sir Thomas Waterlow, Chairman of the Commercial Banking Group, convened a special meeting with the Loganair Board to consider Loganair's first Annual Accounts since the Bank took ownership. This had been followed up three months later when the Royal Bank of Scotland Chairman, Iain McDonald, attended a Loganair Board Meeting and declared bluntly that present losses could not continue. The Bank's Annual Accounts for 1972 may well have stated that the Bank had absorbed Loganair's previous losses because of the airline's social value but, by the time he became Loganair's Chairman, John Burke was under no illusions about his Bank Board's serious concerns about Loganair.

Throughout the eleven years John Burke was Loganair Chairman, he used his considerable energy, his forceful personality and, particularly, the authority that his bank position gave him, to drive forward the seemingly endless dialogue with BEA despite the state airline's prevarication particularly where route rationalisation and the Scottish Ambulance contract were concerned. He constantly argued with the Scottish Office, in all its guises, for financial subsidy for the Third Level air services that Loganair was providing. The Board of the Royal Bank of Scotland, and no doubt its shareholders also, believed that they had been subsidising Loganair's services and Scottish air transport on the understanding they would be recompensed in due course. For his justification, John Burke referred to the CAA Report of March 1974 which had recognised the need for financial support of Scottish air services. Nothing, however, was happening quickly enough, and Loganair was incurring significant losses in 1974/75 made worse by the fuel crisis. John Burke wrote to the Scottish Development Department:

> Barring some immediate solution there will be no alternative to our embarking upon a drastic curtailment of the whole enterprise, possibly going so far as a controlled winding up. I appreciate that this would leave a vacuum in domestic air services in Scotland, which we should all regret, but the shareholders of Loganair are simply not prepared to go on indefinitely subsidising local air transport.

There was much correspondence, and plenty of sympathy, but no positive action from the Scottish Office as there was still no enabling legislation to give financial assistance. The Board of the Royal Bank of Scotland meanwhile resolved to come to a decision, one way or the other, as to the future of Loganair by 31 March 1975, the Company's year end. To the great relief of Loganair's management, financial support was eventually promised under the new enabling powers of Section 21 of the Scottish Development Act which came into operation in December 1975. This was the breakthrough it had been waiting for. It had been a close run thing.

The Company was indeed fortunate to have such an influential and well respected Chairman as John Burke, a man who could open the necessary doors. He also knew, however, that he had given an undertaking to the Board of the Royal Bank 'to eliminate this burden from their Accounts' and as far as the future of Loganair was concerned, he had to keep all his options open. When asked if he would consider disposing of Loganair, John Burke acknowledged the possibility and pointed out that:

> It was rather unusual for a Bank to run aeroplanes, and if an operator more skilled in running an airline were to show an interest, we would consider it. We have had many approaches to buy Loganair but have not had any which we think we could rely on to carry on this operation – virtually an

essential service, certainly in Orkney and Shetland. The possibility of giving it over to someone who might mishandle it is not to be thought of.

John Burke had indeed had expressions of interest from a range of parties but, no doubt reflecting the Bank's position, he felt the great weight of social responsibility and decided that none of the applicants had the necessary, proven experience in flying scheduled air services. Peter M Kaye, who had property and aviation interests, and who owned the island of Little Cumbrae in the Firth of Clyde, was one party who had made an approach in 1972, but had not been given an encouraging response from John Burke. Similarly, an approach by former Loganair pilot, Geoff Rosenbloom of Airgo, and Sir Hugh Fraser were also unenthusiastically received.

The Board of the Royal Bank of Scotland, however, did consider seriously an approach by Sir Kenneth McAlpine of McAlpine Aviation who was planning to base Hawker Siddeley 125 executive jets and several smaller aircraft, all for charter operations, at Glasgow and Inverness airports. The dialogue at Board level would range from McAlpine having a minority stake in Loganair, to a seventy-five percent majority holding, but the Bank Board felt that the deal would be more acceptable to the CAA if the Bank remained as a fifty-one percent majority shareholder. There were protracted discussions but the interest of both companies cooled, McAlpine partly because of the severe impact that the fuel crisis was having on its business, and Loganair because it was now taking the view that the CAA's Report published in April 1974 made it more attractive for Loganair to remain on its own.

By now the Board was also taking greater encouragement from renewed and more genuine enthusiasm on the part of British Airways, which had resulted from the merging of BEA and BOAC, to establish a more meaningful working arrangement. The Board was also no doubt influenced by the fact that the CAA and the Scottish Office had indicated informally that it would be in Loganair's best interests not to have equity links with either McAlpine Aviation or British Airways.

### *Orkney*

In 1974, the Company was continuing to incur worrying financial losses and the relationship with BEA had been frustrating in the extreme, but the air ambulance was now a great success and the inter-island air services in Orkney had also settled down into a most successful operation.

Back on the morning of 27 September 1967, after some irritating delays in delivery of its first Britten-Norman Islander, G-AVKC, and getting the runway lengths of the island airstrips approved by the Board of Trade, Loganair's Orkney North Isles scheduled service operation had got underway. Captain Ken Foster, accompanied by local MP, Jo Grimond, and five passengers, took the first flight from its Kirkwall base to the islands of Stronsay, Sanday and North Ronaldsay. Scheduled services were expanded the following month to Westray and Papa Westray.

The air link between these two islands strips, a distance described by Orkney Captain Andy Alsop as two-thirds of the length of Heathrow's longest runway, and scheduled for two minutes flying 'to allow for a headwind!', would attract international recognition in the years to come through its entry in the *Guinness Book of Records* as 'The Shortest Scheduled Service in the World'.

Later, in 1971, the island of Eday, with its 'London Airport' named after the nearby Bay of London, was added, as was Hoy in 1973. Lastly in 1977, Flotta completed the Orkney Islands network.

Loganair's air services in Orkney were in effect run under contract to state-owned Orkney Islands Shipping Company, a mechanism agreed by the Scottish Office to provide subsidy for Loganair and to protect the Company against operating losses. The OISC was responsible for determining air service frequency and timetable, but there was close collaboration between Allan Bullen, OISC General Manager, and Loganair's pilot management, initially Captain Jim Lee, and then Captain Andy Alsop who would be Senior Pilot, Kirkwall, for some ten years. Relationships between Loganair and the Shipping Company were cordial and constructive and the air service developed and flourished.

NORTH ISLES AIR SERVICE TIMETABLE

The shape of the archipelago was ideal for radial air services linking Kirkwall, Orkney's administrative and commercial centre, with the North Isles airstrips. The boat services historically had been the lifeline for the islanders, but Loganair's Britten-Norman Islander service from 1967 changed all that. The boat took some four and a half hours to reach North Ronaldsay, the most remote of the Orkney Islands. Formerly, having to take the steamer to do business or shop in Kirkwall, the islanders might require to spend one, or even two, nights away from home. The journey by air now took fifteen minutes. Moreover, it was not uncommon for the boat to reach North Ronaldsay and then have to turn around without touching the pier because of impossible sea conditions. The new air service was now the lifeline, even although a survey showed that the islanders thought that the air fares, £2 or £2 10s single to Kirkwall, or £1 single between each island, were too high.

*The Kirkwall team of 1972 under Senior Pilot, Captain Andy Alsop which did so much to establish the Orkney North Isles air services. Left to right: Bryan Sutherland, Ken Milligan, Captain Andy Alsop, Kathleen Matches, Captain Jamie Bayley, Bob Tullock and Brian Anderson.*

Courtesy of Ken Foster

The new air services were hugely popular with the communities of the North Isles. Willie Ross, Secretary of State for Scotland, and the Scottish Office, the subsidising body, also had reason to be pleased at the reduction in Orkney's direct and indirect transport subsidies as, with the air service in place, it had been possible already to replace the old *Earl Sigurd*, a cargo and passenger vessel, with an all-cargo vessel which was some £100,000 cheaper than a replacement cargo and passenger boat.

Passenger usage of the North Isles air services grew strongly year on year. The Kirkwall-based Britten-Norman Islander meanwhile was being deployed increasingly on ambulance missions, not only within the islands, but from Orkney to the mainland, especially Aberdeen and Inverness. Many islanders had reason to be thankful for the commitment of Loganair's staff at Kirkwall. Everything was progressing well when there was rather unwelcome publicity for the Company following a bizarre incident. Captain Jim Lee, Loganair's Senior Pilot in Kirkwall, was dismissed from the Company by Managing Director Captain Duncan McIntosh, based in Glasgow.

It had been Captain Jim Lee who had delivered Islander G-AVKC to Kirkwall in August 1967 and he had been in Orkney throughout the entire two and a half years of the local operation. He was a popular figure utterly committed to the task in hand. His offence was to break into BEA's store at Kirkwall airport and collect parcels of newspapers to be delivered to Shetland. Allegedly this was not the first time he had forced his way into BEA's premises and BEA complained formally. Captain Lee would protest that he was intent on providing a service and it was not an act of hooliganism, but to no avail. He was dismissed. The islanders raised a petition which had a great many signatures, including many from the local medical profession.

Orkney MP, Jo Grimond, tried to intercede but Captain Lee was not reinstated. The whole affair was regrettable. The Royal Bank of Scotland most certainly made it clear to Duncan McIntosh that they did not appreciate national publicity of this kind. Reports that Captain Lee had once landed his Britten-Norman Islander aircraft on one of the Churchill Barriers are assumed to be apocryphal.

## *Shetland*

The Shetland Islands are the most northerly in the UK, some seventy miles from the most northerly tip of Scotland and fifty miles north of the Orkney Islands. Interestingly, the islands occupy a more northerly latitude than either Moscow or the southern tip of Greenland and are closer to Bergen in Norway than to Aberdeen. In midsummer, it never really becomes dark with the sun barely dipping below the horizon, a phenomenon called the 'simmer dim', but severe weather conditions can also make Shetland a most challenging operating environment. Ferry services were well-developed and represented the islands' lifeline. But ferries, especially to the more distant islands of Fair Isle, Foula or Out Skerries, can be stormbound for several days at a time. Loganair's Britten-Norman Islander could surely provide an essential link to communities which could otherwise be cut off.

Once the Orkney air services had been introduced, there was a general expectation that a similar operation in Shetland would soon follow. Zetland County Council had approached the Company in 1967 and Loganair immediately expressed willingness to base a Britten-Norman Islander in Shetland. The Company pointed out, however, that some financial guarantees would be required and, importantly, new airstrips on the outer isles were needed and some work required on the existing grass strip near Lerwick. Loganair was aware that the Shetland geography was different from Orkney where radial air services were possible. Shetland, by comparison, was mainly a long chain of islands with the main airport, Sumburgh, situated in the south, and the newly constructed airstrip at Unst located in the far north. The importance of having intermediate stops, especially at Tingwall, six miles from Lerwick, Shetland's cultural and economic centre, was well understood by Loganair from the outset.

Captain Alan Whitfield was given the responsibility of setting up the Loganair operation in Shetland just as Jim Lee and Andy Alsop had done a couple of years earlier in Orkney. Unlike Orkney, there was no local aviation tradition in Shetland. Captain Ted Fresson and his Highland Airways had been developing air services between the Orkney Islands from 1933. It was 1936 before Eric Gandar Dower's Allied Airways from Aberdeen, and Ted Fresson's Highland Airways from Inverness and Kirkwall, arrived at Sumburgh Airport. It would be more than thirty years later

## Shetland airstrips

**Unst 1968**
*Built under the OPMAC Scheme, redeveloped in 1978 at a cost of £9.5 million.*

**Foula 1969**
*The population of thirty-three built the first airstrip with picks and shovels and a small tractor. First landing in October 1969. Operational June 1971.*

**Fetlar 1972**
*Community effort to raise half the cost, the rest coming from HIDB. Old pre-war whaling and air ambulance funds were used.*

**Whalsay 1972**
*Again, community effort with HIDB help.*

**Papa Stour 1972**
*A herculean effort by islanders equipped with only picks and shovels.*

**Out Skerries 1973**
*The short 1,250 feet strip cost £10,000 of which seventy-five percent was from HIDB with the twenty-five island households raising an amazing £1,277.*

**Fair Isle 1973**
*A new airstrip replaced a World War Two strip and was licensed in 1976.*

**Tingwall 1971**
*The first strip near Lerwick was built on Church of Scotland land, but due to boggy condition closed in 1974. A new tarmac airstrip was opened in 1976.*

that serious consideration was given to providing air services to other points in Shetland. Alan Whitfield began a limited inter-island service between Sumburgh and Unst on 1 September 1969 but it was an inauspicious start with progress frustratingly held up by airfield licensing regulations and planning problems.

For the residents of the outer isles, the new air services offered a lifeline, especially in the form of air ambulance cover. With encouragement and guidance from Captain Whitfield, it was the islanders themselves, and not the Council, who started to prepare airstrips for ambulance and non-scheduled flights. The small population of Foula completed a makeshift airstrip in three weeks with a small tractor and picks and shovels. Sites were also cleared on Fetlar and later on Whalsay, Papa Stour and Out Skerries. The critical airstrip, however, was Tingwall, in relation to which there had been several meetings with Zetland County Council, but no progress. Loganair meanwhile was incurring heavy financial losses in Shetland, having managed only five hours flying in December 1969 and only four hours in February 1970 with the Shetland-based aircraft. Necessary airfield licences had still not been obtained. So, in April 1970, with the Board of Trade, and Highlands and Islands Development Board in its corner, Loganair gave Zetland County Council what was effectively an ultimatum that unless the Council showed willingness to have airstrips licensed within the next month, the Company would have to withdraw from Shetland. Despite lack of progress, the Company did not pull out.

As everyone knew, the key to the scheduled service operation was the airstrip at Tingwall. Eventually, with financial support from HIDB, Loganair decided to take

Photo: Dennis Coutts

*Britten Norman Islander, G-AVRA. Captain Alan Whitfield's first landing on the island of Foula in Shetland on 4 November 1970 was greeted by almost all of the population of thirty-three, many of whom had helped to build the airstrip with picks and shovels. Similar efforts by the islanders in Fetlar, Whalsay, Papa Stour, Out Skerries and Fair Isle allowed Loganair to operate inter-island air services including ambulance flights. After a short period of time, a more suitable landing strip was identified on Foula and is still used today.*

own aircraft and setting up its own airline. The Council would also approach another company – air taxi company Peregrine Air Services had been mentioned – to take over the service. Clearly the relationship with Loganair had reached a low ebb.

During the next month, dialogue and vituperative correspondence continued between the Council's Chief Executive Ian Clark and John Burke. The interruption to the services, however, helped focus minds and more meaningful discussions about subsidy and the opening of a new Tingwall airstrip began. Relations thawed a little in November 1975 when the Council, now Shetland Islands Council (SIC) following Scottish regionalisation, agreed to pay for one charter service per week, starting on 1 December - a temporary agreement was reached with the Council and to everyone's relief the inter-island air services resumed.

Photo: Dave Wheeler

*The islanders of Fair Isle, Fetlar, Whalsay, Papa Stour, Out Skerries and Foula helped in the construction of their airstrips under the guidance of Captain Alan Whitfield. With the aid of a Loganair Islander, Shetland ponies were re-introduced to Fair Isle after an absence of eighty years.*

Loganair agreed to support the Council in obtaining approval for the licensing of Fair Isle airstrip and, more importantly, in getting approval for the construction of the new Tingwall airstrip. To everyone's relief, the Tingwall's 670-metre tarmac runway was completed by Northern Ireland firm Knockbreda in only twelve weeks and was first used by Captain Alan Whitfield in July 1976 for ambulance purposes. The official opening of the new airstrip took place on 20 October 1976. Even then the *Shetland Times* reported that the future of the inter-island air service, and with it the round-the-clock air ambulance service for the outer isles, was still in the balance while the subsidy renewal was being considered by the Council.

As expected, the inter-island service was given a boost by the opening of the 'new' Tingwall Airport. Services had been operating from Sumburgh to Unst on Mondays, Wednesdays and Fridays with stops at Whalsay and Fetlar on Wednesdays. Stops would now be made at Tingwall on three days a week. A major shift in attitude on the part of the Council was noted, but even then there was still reluctance to subsidise the Loganair operation. Eventually, the Company would accept a lower level of subsidy plus guaranteed charter work through to 31 March 1977. Passenger carryings did improve further and there was more encouragement when the Fair Isle service operated through the winter of 1976/77.

There was real concern when it was announced that the Loganair hangar at Sumburgh Airport, needed for the Britten-Norman Islander, was being demolished. Early hopes that SIC, or even the CAA, as owners of Sumburgh Airport, might contribute to the cost of a replacement hangar were quickly dashed and Loganair paid the net cost of £75,000 after a twenty-five percent grant from HIDB. The Company had welcomed the opening of the new Tingwall airstrip, but now requested that SIC license one of the two runways at the former World War Two Scatsta Airport near the newly-built oil terminal of Sullom Voe. SIC acted quickly to do so before Clyde Developers Limited could license the second runway, as one runway would block the other. Scatsta opened in 1978 and Loganair operated two daily round trips between Sumburgh and Scatsta, but passenger traffic never really took off.

In fact passenger traffic on the growing Shetland internal network remained very disappointing and as late as 1980 the services carried only an average of two passengers per flight. Loganair's efforts to promote the air services were made all the more difficult by SIC's decision to provide free travel to islanders on the inter-island ferries. Loganair was losing £70,000 per annum on its Shetland operations and SIC agreed to a contribution of £20,000 to cover the period to March 1981. Beyond that date there was uncertainty and once again Loganair was having to consider withdrawal.

If the inter-island scheduled services were important to Shetland, the locally based air ambulance service was vital. But not vital enough, it seemed, for SIC to pay for the service which, it insisted, 'was free on the mainland.' SIC agreed to pay half the cost of the service, but there was then a very public row between it and the Scottish Office, which for several months contributed nothing so that even George Younger, the Secretary of State for Scotland, got involved. A concession to SIC would be setting a precedent for other islands. Meanwhile, as the author made clear to the press, Loganair was 'pig in the middle' and continued to lose money in Shetland.

The impasse was broken when Loganair applied to the Scottish Ambulance Service for an increase in its pan-Scottish Contract Standing Charge to recognise the cost of the second Shetland-based pilot who was employed only because of the requirement to provide twenty-four-hour ambulance cover at Tingwall. This was agreed in July 1982 and once again the day was saved.

There was another very important development. SIC, with much encouragement from the Chairman of its Transportation Committee, Jim Irvine, awarded to the Company the SIC Oil Pollution Contract which started in January 1982. This required a second Islander aircraft to be based in Shetland and clearly was awarded with a view to helping stabilise Loganair's financial position in Shetland.

### British Airways Route Rationalisation

In April 1974, Lord Boyd-Carpenter, Chairman of the CAA, published his eagerly awaited Report, *Air Transport in the Scottish Highlands and Islands*. The Loganair management was greatly encouraged by the Report's suggestion that 'there should be a competitive tender for the Tiree–Barra route forthwith', which was a clear challenge to the British Airways-dominated status quo. The Report's recommendations also encouraged a mood for future change in British Airways' third level Highlands and Islands air services. As early as May 1974 a monthly Working Party was set up and comprised representatives of British Airways, including David Hyde, General Manager, and Hugh Reid, Marketing Manager of British Airways Scotland, as well as representatives from Loganair, the Civil Aviation Authority and the Scottish Office. This time there was a determination to make real progress with route rationalisation which more properly would separate 'second' and 'third' level air services in Scotland, not only for the benefit of the air passenger, but hopefully also to improve the perennial financial loss situation of both British Airways and Loganair.

British Airways may have been slow to come to the party, and Loganair was incurring heavy losses, but the reputation and goodwill of the Company had been steadily growing. During a debate in the House of Lords in June 1974, the Minister of State at the Scottish Office, Lord Roberts, said, 'The Loganair services make a significant contribution to the economic wellbeing of many of the most remote communities. They show flexibility, resource, an intimate knowledge of local needs, and a realism to respond to them.' A fitting tribute to Duncan McIntosh and his colleagues.

Even with a clear will to implement the recommendations of the CAA Report, British Airways Regional Division - Scottish, as they were now known, had a few issues to address before any route transfers could take place. Firstly there were the two Short Skyliners which were being used on both the Tiree-Barra and Campbeltown and Islay routes. Unsurprisingly, these two aircraft were as ugly, as noisy, and as unpopular with passengers as Loganair's Skyvan had been. By exercising a return clause, the two aircraft, after only four years service, would have to be returned to

Short Brothers. Next, David Hyde, British Airways' Scottish General Manager had to appease British Airways' Glasgow-based engineers who, of course, were losing the maintenance of the two Skyliners. There was no difficulty with the pilots who would simply move on to the HS748 fleet, two of which had just been purchased to be used initially on the Aberdeen–Shetland route.

Perhaps most tricky of all, David Hyde also had to 'sell' the proposal to the British Airways hierarchy in London. Meanwhile, British Airways Helicopters was making overtures to take over the two west coast routes as it, and others in British Airways, were concerned at the intervention of Loganair. Fortunately, Pat Gillibrand, the Director of the Scottish Regional Division, gave the action plan his blessing and his Press Release was a landmark statement: 'This development is significant in that it rationalises the British Airways fleet, with considerable financial savings, and inaugurates an era of co-operation with Loganair.' Nonetheless, David Hyde later acknowledged that he had been severely reprimanded for disposing of the two Skyliner aircraft and transferring routes without proper authority from London. Fortunately for him, and certainly for Loganair also, the British Airways hierarchy by this time was totally immersed in sorting out the BEA/BOAC merger. Even so, route transfers were being proposed for Tiree and Barra only. British Airways' Scottish management decided to retain Campbeltown and Islay and operate the Viscounts, now nearly twenty years old, as this would be more acceptable to the British Airways Board.

The Joint British Airways/Loganair Working Party got down to work and, from 1975, a series of further route transfers took place. The first of these had been the Tiree–Barra route where Loganair started initially on contract to British Airways from September 1974, and on its own account from 1 April 1975. Then the 'Long Island' route between Stornoway and Benbecula was started in October 1975 with financial support from the Western Isles Islands Council and initially with a Britten-Norman Islander service every weekday that was extended to Barra on two days per week. Next, British Airways gave up its Inverness–Wick–Kirkwall service to Loganair on 1 April 1976, but only to be pressurised into returning to the route with a midday Wick–Kirkwall–Wick rotation for the limited period May to November 1976.

Meanwhile, to Loganair's great disappointment, the Scottish Office stated it was content that British Airways should continue with its once a day Viscount service from Glasgow through Campbeltown to Islay. Happily, this turned out to be only a short delay and these routes also were taken over by Loganair on 1 April 1977 with direct services to each destination with a twice daily Trislander service.

And there was more. Having agonised over its very disappointing passenger carryings on its Edinburgh–Inverness service, British Airways eventually decided to

Courtesy of Sandy Matheson

*Trislander, Barra Beach. The Company started services on the 'Long Island', over the
Stornoway–Benbecula–Barra route, with an Islander in October 1975 with subsidy from the Western
Isles Islands Council and in April 1981 upgraded to the Trislander. Left to right with the arriving
Trislander from Glasgow in 1975: Robert Stewart, Chairman Western Isles Health Board, Hugh
Morrison, Councillor Castlebay, Barra, Dougie Morrison, Councillor Benbecula and South Uist,
Donald MacKay, Chairman Transportation Services WIIC, Sandy Matheson, Convener of WIIC,
Roger Haworth, Director of Research & Development WIIC, Donald George MacLeod, Director of
Finance, and Loganair's Flight Ops Director, Captain Ken Foster.*

come off the route and offer it to Loganair who gratefully started with two and later
three daily Islander services in early 1975. Sadly, this was no money spinner. Even in
the winter months when the road journey alternative was more hazardous, passenger
demand continued to be modest and the route, which the Authorities had stated from
the beginning would never receive financial support, was eventually given up in 1980.
Later efforts were made by Loganair, but these were increasingly less likely to be
successful as the A9 road south from Inverness underwent a major upgrade
programme.

Still outstanding at this late hour, was the small matter of subsidy. The Tiree and
Barra services, and also Stornoway–Benbecula–Barra routes, needed subsidy and
Loganair had assumed financial support would be forthcoming. At last, in January
1975, new legislation, the long awaited Scottish Development Agency Act 1975, made
clear that powers and responsibility for these lifeline services now lay with the local
authority and not HIDB. This was a great relief to everyone and not least of all to John
Burke and the Royal Bank Board who had expressed publicly their impatience and

frustration at the lack of subsidy provision. The Tiree–Barra route, fortunately and conveniently, was a through service linking two regional authority areas and as such was the responsibility of the Scottish Office and not of Strathclyde Regional Council nor of Western Isles Islands Council. As far as Stornoway-Benbecula-Barra – the 'Long Island' service - was concerned, the route fell within the responsibility of the Western Isles Islands Council. Similarly the Inverness–Wick route, wholly within the Highland Region, would be subsidised by Highland Regional Council.

The Long Island service had first been mooted by Highlands and Islands Development Board as far back as 1973. Transport links from Stornoway in Lewis, the administrative centre of the Outer Hebrides, to the southern isles of North Uist, Benbecula, South Uist and Barra were difficult. The Sound of Harris was shallow and difficult to navigate and did not have a vehicular ferry until 1997. Passengers had to travel to Lochmaddy in North Uist by ferry. A direct air service from Stornoway to Benbecula would cut a very long journey to barely an hour. Not surprisingly therefore, the Stornoway–Benbecula service went from strength to strength. Service frequency was increased from once daily to three times daily with the Islander, and later twice daily with the sixteen-seat Trislander, before being replaced by the eighteen-seat de Havilland Twin Otter (and later in the 1980s by the thirty-six seat Shorts 360). The double daily Stornoway–Benbecula service was extended each weekday to operate between Benbecula and Barra.

Taking over the Campbeltown route next, from British Airways on 1 April 1977, turned out to be the easy bit for Loganair. The Company's new twice daily Trislander service allowed a good day's business at both ends of the route which had not been possible when British Airways operated the once daily Viscount. Loganair's passenger carryings immediately showed significant growth over previous years. Unfortunately, despite this early encouragement, Loganair, like British Airways, could not make a profit on the service and, before the end of the first year on the route, the Company decided to approach Strathclyde Regional Council (SRC) for a subsidy. The Scottish Office had made it clear that internal air services which operated within a region were the responsibility of that Regional Authority. For the Glasgow–Campbeltown route this meant SRC.

SRC, for its part, was adamant that, just as the Scottish Office subsidised ferry services, so it should subsidise air services. There had been no history or precedent for SRC to become involved in air services. At one meeting on transport in Glasgow, a Glasgow Councillor expressed the view that the air service was 'for landowners and retired Colonels.' He was clearly out of sympathy and oblivious to the importance to the remoter communities of a quick link to the mainland whether it be for family, business or hospital reasons. SRC was not prepared to subsidise, stating it was the responsibility of the Secretary of State. At least it was consistent, for at this time too, SRC refused to help Burnthills Aviation with its Highland Helicopter Service that was previously subsidised by the HIDB and which provided services from Glasgow to Rothesay, Lochgilphead, Oban and Fort William. Loganair cut frequency in order to save costs and reduce losses and continued operating on a marginal basis for many years until 1997 when the Scottish Office granted the route Public Service Obligation (PSO) status and thereby financial support for the route.

It had not all been successful and it had been a long time in coming, but the Working Party had achieved a lot in a short space of time. Not only had the route transfers been handled well and been successfully undertaken, but there had also been sensible efforts at co-ordinating the British Airways and Loganair timetables, with cross referencing and connecting services highlighted, which greatly benefited the passenger.

### New Members of the Family

Part of David Hyde's concern about these route transfers was their effect on the British Airways staff. There had been no reaction from the pilots as they would simply move from the Skyliners onto the new HS748 aircraft, but he was concerned for the ground staff at the Campbeltown, Islay, Tiree and Barra outstations. British Airways had several managers in Campbeltown and Islay who moved on within the British Airways organisation. In Tiree and Barra, however, although given the opportunity to move, the two British Airways managers resolutely wished to stay where they were.

Tiree's Station Manager from the BEA days was Archie MacArthur, a native of Tiree, the most westerly of the Inner Hebrides. An aircraft first landed on Tiree on 18 July 1929, but it was 1 July 1936 when the island started to receive fairly regular passenger flights, these being inaugurated on 1 July 1936 by Captain David Barclay on behalf of Northern and Scottish Airways. Barclay's first landing on Tiree, was on the beach fringing Gott Bay on 4 October 1935, but during World War Two a substantial airfield with three hard runways was laid down by the RAF for use by Coastal Command, and although only one runway is now in use, it is that airport that continues to serve Tiree today. After the war, BEA served Tiree with Rapides, and later with Herons, although there was a short period in summer 1946 when a Dakota linked Tiree not only with Renfrew but also with Benbecula, Stornoway and Inverness.

Archie MacArthur was in charge at Reef Aerodrome when the Herons and, for a brief period the Skyliners, were providing the air service to the island. He was a gentle and caring man and would become a great ambassador for Loganair. At the time of his transfer to Loganair's employment, he wrote to David Hyde: 'I may say that while I remain at Tiree, I trust Loganair will find that I shall always endeavour to serve them

*It was the Company's practice from the 1970s to name its Britten-Norman Islanders after individuals who had made a contribution to the air ambulance service including many of the great pioneers of Scottish aviation.*

Photo: Iain Hutchison

**Captain E E Fresson OBE** *the founder of Highland Airways who in 1933 started mail and passenger services from Inverness to Orkney and later Shetland and the Western Isles.*

**Captain David Barclay MBE***, a legend among British European Airways pilots, and before that with Midland & Scottish Air Ferries, Northern & Scottish Airways and Scottish Airways.*

**Sir James Young Simpson** *after whom BEA had previously named a de Havilland Heron. Simpson (1811-1870) pioneered the use of anaesthesia to assist child-birth.*

**Robert McKean OBE FCIT***, Chairman of BEA Scottish Airways, had a particular interest in the air ambulance service and was most helpful to Loganair when it took over the contract in 1973.*

**Sister Jean Kennedy** *had already undertaken more than two hundred ambulance flights when she was killed on a BEA ambulance mission to Islay on 28 September 1957. The Heron crashed a mile short of Islay Airport.*

**E L Gandar Dower Esq** *established Aberdeen Airport and set up Aberdeen Airways in 1934, later changing the name to Allied Airways (Gandar Dower) Ltd. His airline provided many air services throughout Scotland, including services to Orkney and later to Shetland.*

**Captain Eric A Starling FRMetS** *was Aberdeen Airways' chief pilot and later served as BEA's flight manager in Scotland. He was heavily involved with Scottish air ambulance flying until retiring in 1971.*

well and the public for whom they cater.' Archie MacArthur certainly was as good as his word and was hugely respected until he retired in 1990. The family tradition continues as his daughter Tish joined the Company in 1988 and succeeded her father on his retiral. Archie was the airline in Tiree. The author remembers attending a public meeting in Tiree in 1976 when reference from the audience was made to BEA and he had the temerity to point out respectfully that it was in fact Loganair, and not BEA who had been operating the Tiree service for the last year or so. To which came the response, 'No, it is BEA – Big Erchie's Airways – and it will always be so.' And quite right too.

As far as neighbouring Barra was concerned, it was Northern and Scottish Airways and its Chief Pilot, David Barclay, which introduced regular flights in 1936 with Spartan Cruisers and de Havilland Rapides. The airline's brochure that year advised intending passengers for Barra to contact 'John MacPherson, Post Office, North Bay'. The following year, British Airways Ltd in conjunction with the London Midland and Scottish Railway and David MacBrayne Ltd, merged Northern and Scottish Airways with Inverness-based Highland Airways to form Scottish Airways. Dragon Rapides continued on the Barra route until 1955, but in 1947, Scottish Airways was absorbed into BEA, which continued serving the island until 1974 when it was merged with BOAC to become British Airways.

David Hyde was also apprehensive about how Katie MacPherson MBE, British Airways' Station Superintendent in Barra, would react to leaving its employment after a lifetime of service. He hired a Cherokee from the Glasgow Flying Club and flew out to Barra on a Saturday to explain the detail of her severance terms to her. She listened, smiled and then declared: 'I did not understand a word, but I trust you, so let's go inside and have a dram.' Now that's the right way to do business!

Katie wished to stay on in Barra and transfer to Loganair's employment thus allowing her to continue a remarkable association of four decades with the beach airport until ill health forced her retirement in 1980. She had joined her father John

Courtesy of Oban Times

*John Burke, Duncan McIntosh and Sir Michael Herries at the opening of Barra terminal building in 1978 and replaced the old BEA hut that had been brought from London's Northolt Airport to Barra after World War Two. Loganair's terminal was formally opened by Sir Michael Herries, Chairman of the Royal Bank of Scotland. Loganair's Chairman, John Burke, was also Managing Director of the Royal Bank of Scotland and Deputy Chairman.*

(the Coddy) MacPherson at the airport in 1936, then worked with her brother Angus after World War Two when he took over from his father as Station Superintendent for BEA, and finally became Station Superintendent herself in 1950 for the next thirty years. This was a remarkable MacPherson family association which was to be extended when Katie was succeeded in Loganair's Barra office by her niece, Una MacPherson.

The duties and responsibilities of the Barra Station Superintendent in these earlier years were varied and demanding. Landing an aircraft on a cockleshell beach requires considerable skill and the pilots came to depend on Katie MacPherson's experience and judgement in giving her assessment of weather and landing conditions, which she did on a hand-held radio, to the approaching aircraft. Little wonder, then, that Katie MacPherson was awarded an MBE in the Queen's Honours List in 1969 for her contribution to the Barra air service. Later this shy and modest lady earned further recognition with the Woman of the Year Award.

### *The Aircraft of Choice – the Britten-Norman Trislander*
The Company's earlier, damaging decision to acquire the Beech 18, and more especially the Shorts Skyvan, had shown the importance of choosing the appropriate aircraft for the task in hand. In 1967, Loganair had taken delivery of the first of their fleet of Britten-Norman Islander aircraft. The eight-seat aircraft had proved ideal for air ambulance flying and the short sector, island hopping operations in Orkney and Shetland.

The Britten-Norman Islander had been an excellent choice for Loganair. There had been serious cause for concern in 1971, however, when Britten-Norman, the manufacturer of the aircraft in Bembridge, Isle of Wight, went into receivership. Fortunately it was acquired by the Fairey Group which, in 1977 was taken over by Pilatus of Switzerland and production continued in Gosselies in Belgium. Encouragingly, Britten-Norman was now manufacturing a Trislander, a stretched version of the Islander carrying sixteen passengers with two pilots, and with a third engine mounted high in the tail. Believing that Britten-Norman's future was secure as part of the Fairey Group, Loganair ordered the new Trislander and G-BAXD was delivered in 1973. By buying the bigger, sixteen-passenger seat Britten-Norman Trislander aircraft, Loganair made this route transfer process feasible and had themselves become more credible successors to British Airways.

When the Edwards Committee and CAA reports recognised the distinction between second and third level air services, and Loganair's discussions with British Airways eventually were starting to make some progress, the sixteen-seat Britten-Norman Trislander had come into the Company's reckoning. The Company eventually bought

*For nearly thirty of her forty-four years employed at Barra's unique beach airfield until her retiral in 1980, Katie MacPherson was Station Manager. Her responsibilities were varied and included passing to the pilots overhead, on her hand-held radio, her assessment of weather and landing conditions. For her outstanding contribution to Barra's air services, she was Woman of the Year and in 1969 she was awarded the MBE.*

Photo: British Airways

eight Trislanders in all to make it possible for Loganair to be taken seriously as the new operator of British Airways routes. Indeed, when the Glasgow–Campbeltown, Glasgow-Islay and the Edinburgh–Inverness routes were transferred, the Company could operate a higher service frequency because the Trislanders' economics and seating capacity were manifestly more appropriate than British Airways' seventy-two seat Viscounts.

There was a natural preoccupation with developing a more robust scheduled service network. However, there was also a growing awareness of the commercial opportunities afforded by the explosive expansion of North Sea oil exploration. The Company had the Trislander and used it during the very early days of the North Sea oil boom. The Trislander, however, very quickly became unpopular with the burgeoning North Sea Oil industry. It never became a successful aircraft. While some 1,280 Britten-Norman Islanders were to be manufactured over a forty-year period, the Trislander production stopped within less than ten years and with only seventy-two aircraft manufactured. Loganair's eight Trislanders were bought between 1973 and 1977, but were no longer in the fleet by 1982. Seven were sold on and one was written off in a heavy landing at Aberdeen Airport in 1979.

In a very competitive market place, Loganair's Trislanders were losing out to the de Havilland DHC6 Twin Otter, and later, to the Brazilian Embraer EMB110 Bandeirante. If it was going to compete, Loganair had to make investment in new aircraft. It opted for the eighteen-seat de Havilland DH6 Twin Otter which was generally regarded as an excellent workhorse to support the oil industry. In particular, it would be a suitable aircraft to secure work at Occidental's huge oil terminal on the island of Flotta in Scapa Flow in Orkney, and Chevron's operation on the island of Unst in Shetland. It was quicker, quieter and much more comfortable than the Trislander and, most importantly, it had an excellent short take-off and landing (STOL) capability.

*Britten-Norman Trislander, G-BBNL. Loganair acquired the first of its eight, sixteen-passenger Trislanders in 1973 and it first appeared on the Dundee–Glasgow route. The Trislander allowed the Company successfully to take over the Glasgow–Campbeltown and Glasgow–Islay services, replacing British European Airways' Viscounts. BEA had also planned to acquire the Trislander but became concerned at the financial difficulties being experienced by its manufacturer, Britten-Norman Limited of the Isle of Wight. This 'big brother' of the successful eight-seat Islander never became popular. Production ceased within ten years with only seventy-two aircraft manufactured. Loganair disposed of its last Trislander in 1982 when it was replaced by the larger, de Havilland Twin Otter on the Stornoway-Benbecula-Barra route.*

### The Embraer 110 Bandeirante

The Board saw one further opportunity. By 1980, it had become clear that the Twin Otters were not meeting all the requirements of the oil industry and business was being lost to operators of the Brazilian-built Embraer 110 Bandeirante. BP and BNOC had recently stipulated Bandeirante aircraft for their contracts. The Loganair Board was aware that the Brazilian Government was supporting the export of these aircraft by making very cheap finance available through Banco Real. Banco Real would provide a US Dollar Loan for the purchase price of the aircraft – A Forward Contract at the very attractive exchange rate of $2.55/£ was also arranged by Loganair – capital repayments were to be over seven and a half years at six monthly intervals at seven and a half percent interest on the reducing balance. All this was at a time when UK bank interest rates were around sixteen percent. Little wonder then that for the first time, the Royal Bank was not involved in financing Loganair aircraft acquisitions. Little wonder also that Loganair kept the Banco Real financial arrangements in place long after its two Bandeirantes had been sold. In an increasingly competitive North Sea oil charter market in Aberdeen, having the Bandeirante with its greater speed and range, and indeed passenger comfort, did allow Loganair to win work from many oil companies.

David Dyer Collection

*The Company bought its two Embraer 110 Bandeirantes in 1980 having taken advantage of a particularly attractive funding package from the Brazilian Government. The two eighteen-seat, unpressurised aircraft were employed primarily on North Sea oil charter work. The Company's Bandeirante operations were relatively short-lived. The first aircraft was sold in 1984 to Provincetown-Boston Airline and the second in 1986 to Jersey European Airways.*

### And getting them here

These new aircraft, the Twin Otter and Bandeirante, were manufactured in Canada and Brazil respectively. Loganair chose to undertake the delivery ferry flights and there was no shortage of intrepid pilots volunteering for the quite dramatic change from the short hops to which they were more accustomed. Most of the Twin Otters were collected from de Havilland Canada at Toronto and a typical delivery route would take the aircraft, fitted with ten forty-five-gallon overload tanks, to Sept-Iles in Quebec and

Goose Bay in Newfoundland before they set out across the Atlantic. A stop would usually be made at Narssarssuak in Greenland although Captain Ken Foster recalls one delivery where this stop was omitted and touchdown was made at the next port of call, Keflavik in Iceland, and still with sufficient fuel to take them on to Glasgow.

The longest Twin Otter delivery was that of G-BHFD undertaken by Captains Alan Whitfield and Ken Dempster from Fields Aviation at Calgary and firstly they flew to Churchill on the shores of Hudson Bay. It was December and the next port of call was to have been Frobisher Bay on Baffin Island, but because of a snow blizzard there, an intermediate stop at Coral Harbour on Southampton Island at the entrance to Hudson Bay was added. From Frobisher Bay 'Foxtrot Delta' continued to Sondre Stromfjord in Greenland, thence to Keflavik and on from there to reach Glasgow after a total time in flight of eighteen hours.

The furthest journey made during any delivery flight, however, was that made by Captain Jamie Bayley in delivering Embraer Bandeirante G-BIBE from the manufacturers in Brazil. Captain Bayley's account of the flight home is so matter-of-fact and captures the spirit and ethos of Loganair that it is worth recalling:

> On 28th October 1980, myself and John White (Flight Administration Officer and Navigator) left São José dos Campos for Recife just south of São Paulo for customs clearance and formalities before proceeding to Fernando de Noronha, a small island in the South Atlantic which is part of Brazil and some three hundred and fifty nautical miles to the north east of Recife. From Fernando do Noronha our route was to Dakar in West Africa, Las Palmas, Cardiff, Glasgow.
>
> The aircraft carried a one thousand litre ferry tank and a portable Omega navigation system and portable HF set in addition to the normal avionics fit. The HF set did not ever function satisfactorily, but the Omega set was of great assistance and performed well throughout the trip.
>
> From São José to Recife took seven hours flight time in excellent weather.
>
> The business of clearing customs at Recife was both protracted and frustrating. Despite the fact that the aircraft was manufactured in Brazil and all the documentation was in order, it took some three hours to clear Recife. I had been hoping not to arrive at Fernando de Noronha during the hours of darkness as the only facility there is an NDB approach and scrutiny of the approach charts showed some very interesting spot heights in close proximity to the airfield.
>
> In the event we arrived in darkness.
>
> On establishing contact with the air traffic controller it was a surprise

to discover he did not speak English (English being the international language employed in aviation).

The night was spent in a Nissen Hut belonging to the Brazilian Army. It was in a state of disrepair and the toilet facilities are best left to the imagination.

We visited the meteorological office which was a small hut on the airfield boundary at 0700 on the morning of the 29th. The meteorological officer (?) cleared a roosting chicken from his desk and produced our route forecast – a transcript of yesterday's weather at Dakar. Communications between Recife and Fernando had now broken down 'but may be working later, Senhor'.

So, armed with the previous day's newspaper, which included a satellite photograph of the South Atlantic weather pattern (it looked fine), we set off and duly arrived five minutes early at Dakar after an elapsed flight time of seven hours thirty minutes.

The remainder of the flight was mainly mundane and hardly worthy of expansion.

The routing was through Las Palmas–Cardiff–Glasgow arriving on the evening of the 30th October.

'Mundane' was not the word most would use. Everyone in the Company was relieved when Jamie Bayley and John White arrived back safe and sound at Glasgow.


### Oil Support Work

To add insult to injury, only a short period after Loganair had discontinued its ambitious, but financially disastrous Glasgow–Aberdeen–Stavanger scheduled service, oil was discovered in the North Sea. Airlines like Air Anglia and Scandinavian Airlines System (SAS) would benefit hugely from North Sea oil activity, but sadly it was too late for Loganair's Stavanger service. There is no doubt that John Burke, and particularly the Royal Bank's Business Development Department in Aberdeen, were actively enthusiastic for Loganair to become involved in the oil business.

If Loganair was to be a player and compete in the Aberdeen oil support market place, the Company would now require to have appropriate aircraft. Initially the Company tested the water with the available Trislander which fortuitously was the right size for crew change operations with the North Sea helicopter work horse, the Sikorsky S61. The Company's first involvement in oil support work was a Shell contract in 1973 to fly oil workers from Aberdeen to Sumburgh Airport in Shetland where they transferred on to a helicopter to be taken to the North Sea oil rigs in the

East Shetland basin. Loganair based its first Trislander at Sumburgh Airport to take advantage of this new business and it was there for a year before the number of Shell personnel had increased to such an extent that Dan-Air's forty-four seat HS748 aircraft took over the contract.

Aberdeen was developing as an important oil centre and Loganair moved its base of oil operations to Dyce Airport with two Trislanders. This was a period of intense oilfield exploration which acted as a honey pot so that a number of airlines and aviation companies saw their opportunity and also quickly established themselves in Aberdeen. There was a lot of work available in the shape of ad hoc charters and short-term contracts of three to six months duration, but competition for each was keen. There had been some success with all the major oil companies including Occidental, Union Oil, Texaco, Total and Elf, but the biggest breakthrough came from Burmah Oil (subsequently British National Oil Corporation) for a two Trislander contract for a period of three years. Although some oil-related flying involved flights to Ireland, Wales or indeed Norway, the 'bread and butter' flying was between Aberdeen and Sumburgh, which often involved long waits for the helicopter passengers at Sumburgh and meant that possibly only three or four hours flying was achievable in a day, perpetuating the age old Scottish aviation problem of getting adequate aircraft utilisation.

Occidental had developed a huge oil base on the island of Flotta in Scapa Flow in Orkney. Chevron was looking to use the airport at Unst, the most northerly of the Shetland Islands to support its Ninian field. The Twin Otter could perform admirably into both airfields. Early expectations of work with Occidental fell through at the eleventh hour, but gradually contract work for Chevron was secured flying directly to Unst which was closest to its Ninian oil field. Having the Twin Otter did indeed secure

*There was great excitement at the delivery of Loganair's first de Havilland Twin Otter, G-BELS, in March 1977 – especially for the author's twin sons Christopher and Geoffrey who borrowed the hats of Captain McIntosh and Captain Jamie Bayley for the occasion.*

Courtesy of Loganair

the Chevron business for Loganair. Better still, Chevron's transfer requirements increased from one Twin Otter to three and then to six with a seventh Twin Otter as backup. Loganair bought or leased the additional Twin Otters for the three-year contract.

Clearly this was Loganair's most significant contract to date. The Loganair team at Aberdeen, by now under Captain Don Scobbie and including pilots, engineers and commercial and administrative staff, worked tirelessly to ensure that the challenging flying operation to Unst went smoothly. However, an event which would have serious consequences for Loganair took place in May 1979. De Havilland of Canada demonstrated its fifty-seat Dash 7 aircraft to the oil industry and took a party of representatives from the Aberdeen oil companies on a demonstration flight - comfortably landing the aircraft on Unst's 2,100ft runway and even landing safely at the shorter Tingwall Airport near Lerwick. The bigger cabin of the Dash 7, and most particularly its exceptional STOL capability, clearly impressed the oil executives.

It took only a short time after that for Chevron to decide to foreclose on Loganair's three-year contract and invite tenders for a new contract specifying the Dash 7 aircraft.

Courtesy of Gilbert Fraser

*Senior Pilot weekend meetings were held twice a year at Glasgow in the 1970s. The Senior Pilot from each outstation had responsibility for 'managing' his base as well as carrying out flying duties. Left to right: Capt Ken Foster, Director Flight Operations; Capt Alan Whitfield, Senior Pilot–Shetland; Capt Andy Alsop, Senior Pilot–Orkney; the author, Finance Director; Capt Duncan McIntosh, Managing Director; Capt Alistair Wallace, Passenger Services Manager; Capt 'Barney' Barron, Senior Pilot–Inverness; Capt Keith Alderson, Senior Pilot–Aberdeen; Gilbert Fraser, Engineering Director; Captain Bill Henley, Chief Training Captain.*

The Company had thought it was servicing a niche market at Unst and had invested in the purchase of six Twin Otters, each costing around £½ million. The Loganair management learned a very hard lesson. A three-year contract, or even a thirty-year contract, with a three-month break clause, in effect was a three-month contract. Negotiations were started immediately with de Havilland of Canada in Toronto for a Dash 7 to enable the Company to submit a tender to Chevron, but the new contract was awarded to Brymon Airways based in Plymouth.

The Company made one last throw of the dice. The Embraer 110 Bandeirante with its greater speed and range was winning more and more oil support work to the detriment of the Twin Otter. The Board took advantage of very attractive financing arrangements to buy two Bandeirantes and new work was indeed won from BP to fly from Aberdeen to Kristiansund in Norway, from Zapata to fly to Amsterdam, from Gulf Oil to Bergen and for Phillips to Norwich. But it was never enough.

The staff at Loganair's Aberdeen base had worked tirelessly to secure market share and, having won the Chevron contract, they had every reason to believe their efforts had been successful. It was to their great credit that the Unst operation performed well until Chevron terminated the contract with three months notice. Losing the Chevron contract marked the beginning of the end of Loganair's activities, and indeed aspirations, in the highly competitive North Sea oil business. More than that, it almost dealt a mortal blow to the Company itself, now having a fleet of Twin Otters suddenly surplus to requirements.

It had been difficult throughout with margins always being squeezed. Loganair, for most of its time in Aberdeen, had other severe problems through pilot shortage. Indeed, at one stage to help recruit pilots, Loganair found it necessary to buy two apartments in Aberdeen and provide accommodation.

After the loss of the Chevron contract, the decision was taken to reduce the Aberdeen oil charter fleet and to put up for sale the hangar at Dyce Airport, a little more than a year after it had been opened. It was leased to many parties, including British Airways Helicopters, Shell and Eagle Aircraft, but it was December 1988 before it was eventually sold to British Airways/Brymon.

For Loganair, and indeed many other airlines that opened bases at Aberdeen, the benefits from the North Sea oil boom had been short-lived. A bitter-sweet experience.

### The love-hate relationship with British Airways

Meanwhile, by 1978, there was a feeling that the British Airways/Loganair Planning Group had probably run its course, but in fact it would continue for a further three years. The aviation scene in Scotland was changing. Most significantly there was another credible third-level operator in the shape of Air Ecosse, the Aberdeen-based subsidiary of English charter company, Fairflight. This caused the CAA to withdraw from the Planning Group, explaining that its original purpose was to implement the CAA Report of 1974 and that programme was now complete. Air Ecosse operated a fleet of Bandeirante aircraft from their Aberdeen base. They had a number of night mail contracts for Datapost and were aggressively seeking daytime scheduled services.

The relationship between British Airways and Loganair at this time was frequently strained but endured through mutual self interest and pragmatism for, on some issues, they had common purpose. By this time, many interested parties had formed the clear view that the dual supremacy of British Airways and Loganair on the Highlands and Islands air routes was just too cosy. When Air Ecosse applied for scheduled service licences on British Airways' Aberdeen–Wick–Shetland route, and was expected also to apply for the Aberdeen–Stornoway route licence, British Airways and Loganair collaborated in an effort to pre-empt it. Air Ecosse was proposing a three times daily service with the eighteen-seat Bandeirante instead of British Airways' once daily service with the forty-four seat HS 748. Critically, Air Ecosse's licence application had wide support, including from Highland Regional Council and the United Kingdom Atomic Energy Authority (UKAEA) – not least of all because the proposed service would allow the Caithness traveller a visit to Aberdeen and return the same day instead of having to spend one or even two nights in Aberdeen.

For some time, there had been mounting irritation with what many regarded as British Airways' record of frequent fare increases and inconvenient timetables. For its part, British Airways had assumed its traditional moral high ground: 'It is essential that any future development of air services in Scotland should be at British Airways' instigation, and not the result of licence applications from speculators such as Air Ecosse whose only interest is to cream off the profitable routes in Scotland.' British Airways' irritation with Air Ecosse was not eased while it investigated the suspected involvement of one of their long haul pilots in the management of Air Ecosse. In any event, Reg Mulligan, Managing Director of Air Ecosse, left British Airways' employment shortly thereafter.

During this period, Loganair continued to benefit from the special relationship with British Airways. At the request of Shetland Islands Council, Loganair applied for a licence to operate from Edinburgh to Lerwick in Shetland with the Twin Otter. British Airways' standard practice would be to lodge an objection. To overcome British Airways' sensitivities, and also to forestall other competition, particularly from Dan-Air, Loganair agreed to operate the route as a joint venture with British Airways. Profits and losses would be shared equally after all of Loganair's costs for running the service, including, at Loganair's insistence, a Return on Capital of eleven percent

(British Airways' own target rate of return) had been charged in respect of the capital cost of Loganair's Twin Otter. Despite the long two-hour flight, often longer when there was a headwind, in the cramped noisy Twin Otter cabin, the service was very popular as passengers seemed to prefer this non-stop service to British Airways' indirect services via Aberdeen or via Inverness and Kirkwall. The arrangement worked well for both companies.

Having this closer relationship still in place also enabled Loganair to take over the Edinburgh–Belfast route from British Airways, this time partly to pre-empt the aspirations of Air UK, newly formed from the merger of Air Anglia and British Island Airways. This was a huge step forward for Loganair. Later, when British Airways discontinued its evening Glasgow–Inverness service and its morning Inverness–Glasgow service, Loganair stepped into the breach and agreed to overnight a Twin Otter in Inverness and operate the vacated service while British Airways continued to provide the reciprocal services on the route.

Loganair had been granted a licence to operate Edinburgh–Kirkwall following British Airways' change of schedule through Aberdeen. Very neatly, a once daily Shorts 330 service was operated from April 1980 between the double daily Shorts 330 flights on the Edinburgh–Belfast route – better aircraft utilisation at last.

Loganair's Shorts 330 also benefited from a significant contract with British Airways. Because of 'The Troubles' in Northern Ireland, and a mortar attack on Aldergrove Airport in 1974, British Airways crews on the Belfast–London Heathrow Shuttle flights would not night-stop in Belfast, and the forty-two crew members were flown each evening to Glasgow, latterly in a Trident deployed for the purpose. From April 1981, the British Airways crews agreed to fly between Belfast and Glasgow in Loganair's Shorts 330 each night at 2210, and at 0600 each morning. This allowed British Airways to make savings of some £350,000 per annum, and allowed Loganair a most welcome additional income and utilisation of the Shorts 330. On the Glasgow–Belfast route, Loganair operated the antisocially timed flights primarily for the British Airways crews, but when British Airways stopped its lunchtime flight due to lack of Viscount capacity, Loganair stepped in to operate a midday flight. British Airways continued to operate its peak-time morning and evening flights. The British Airways crew transfer contract continued until early 1983.

There were other developments, however, which were less straightforward and which strained the relationship with British Airways almost to breaking point. The Western Isles Islands Council had been dissatisfied with the British Airways service for some time. The Council had invited other operators – the Loganair Board believed at the suggestion of the CAA – to apply for a licence to provide alternative services to British Airways on the Glasgow–Benbecula, Glasgow–Stornoway, and Inverness–

Stornoway routes. Roger Haworth, the Council's Director of Planning and Development, approached not only Loganair, but also Dan-Air, British Midland (BMA), British Island Airways (BIA) and Air Anglia, making clear that the Council's motivation and concern was the level of air fares, rather than timetable improvement.

When, in March 1978, Roger Haworth advised the Loganair board members that another airline was bidding for the Glasgow–Benbecula and Glasgow–Stornoway route licences, it was seriously exercised and resolved to submit its own application for the route licence. For this route application to be taken seriously, the Loganair Board realised that a bigger aircraft was required and the thirty-seat Shorts 330 was immediately evaluated for the Scottish routes. This would be a significant investment for the Company, but the Board took encouragement from their conviction that the British Airways scheduled service scene might be changing again.

Indeed, by the time Short Brothers of Belfast demonstrated its new aircraft in Scotland in September 1979, British Airways had announced its intention to withdraw from twenty-six domestic routes by March 1980. Clearly attitudes were changing within British Airways. Serious financial losses were still being incurred on its Scottish services, the Western Isles Islands Council had demonstrated its discontent with British Airways, and other operators like Alidair, Dan-air and Air Anglia had already been nipping at British Airways' heels by applying, albeit unsuccessfully, for licences to operate from various mainland points to Shetland. Times certainly were a-changing. Loganair had to be properly equipped to compete and take advantage of any withdrawal of services by British Airways – even although British Airways was still robustly refuting any suggestions that it would do so. Loganair placed an order for two Shorts 330 aircraft.

This was clearly not in the spirit of the collaboration of the Planning Group and not likely to please British Airways. The proposed use of the Shorts 330 aircraft by Loganair was breaching the basic plank of the Planning Group whereby Loganair would remain as a Third Level operator. Even more provocatively, Loganair intimated that as well as applying for the Glasgow–Benbecula route licence, it would also be applying for revocation of the route licence held by British Airways.

### Translink

An approach from the British Airports Authority (BAA), which owned Prestwick Airport as well as Glasgow, Edinburgh and Aberdeen, for Loganair to quote for feeder services for Prestwick's Transatlantic scheduled services, had given the Loganair Board further encouragement to acquire the Shorts 330 aircraft. BAA accepted Loganair's tender, despite Air Ecosse having offered Bandeirante services with no subsidy.

*Loganair was the first operator of the Shorts 330 manufactured by Short Brothers of Belfast as a development of their earlier Skyvan and Skyliner aircraft. The Company owned and operated two aircraft between 1979 and 1984.*

*The Shorts 330 had the unfortunate habit of yawing – of snaking in turbulence, until a yaw damper was fitted later. It earned the unenviable name 'Vomit Comet', especially by passengers and cabin staff who were seated right at the rear of the cabin.*

*Loganair used the Shorts 330 initially for the short-lived Translink service on behalf of BAA to provide air links between its Aberdeen and Edinburgh Airports and the international services at Prestwick Airport – Scotland's Transatlantic 'Gateway'. More importantly, it was used extensively on Loganair's new scheduled services including the all important Edinburgh–Belfast service in 1980 and the Edinburgh–Manchester service in 1982.*

Loganair's Shorts 330 Translink Services, as they were called, commenced in July 1979 with morning services from Prestwick to Edinburgh and return, followed by Prestwick–Edinburgh–Aberdeen and return. All too soon, any thoughts that BAA, or indeed Loganair, may have had, that Translink would run like clockwork and be as successful as BAA's London Heathrow–Gatwick Helicopter link were very quickly dispelled. Making connection with a reducing number of Transatlantic flight arrivals at Prestwick, some of which were frequently subject to delays, made Translink a difficult operation. Passenger uptake was disappointing. To make matters worse, Loganair had assumed full sector fares would apply for interline passengers and British Airways offered seventy-five percent.

In any event, it soon became clear that the Scotland–North America market would only support low yield traffic. Loganair's full fare add-on was decidedly unattractive for the market and passengers very quickly found that it was cheaper to fly from North

America to Scotland via London. With the seasonal drop in Prestwick's Transatlantic services, Translink was suspended after barely five months. The increased summer Transatlantic flying programme encouraged BAA and Loganair to restart at the end of April 1980 with a mixture of Twin Otter and Shorts 330 flights.

Translink was a financial failure for Loganair. Even with the BAA subsidy of £160,000 per annum, it was not going to be sustainable with the low passenger demand and low revenue return. Air Ecosse took over the Translink service for a short period with the Bandeirante, but were no more successful than Loganair had been. The age-old problem of adequate aircraft utilisation again reared its ugly head. Translink services, because of the arrival and departure times of the Transatlantic flights at Prestwick, were essentially over by early afternoon. Finding an appropriate task for the Shorts 330 for the rest of the day was critical, but seldom achieved.

### Other Route Licences

British Airways may not have lodged any objection when Loganair had applied for the Translink route licences, but even during the period of the Planning Group, it was routine practice for it to object to licence applications from Loganair or any other operator which might have a direct or even indirect implication for one of its routes.

ASH AWARD

SCOTTISH COMMITTEE OF ACTION ON SMOKING AND HEALTH

*This is to certify that*

*Loganair*

*has been selected for the Ash Award for Services to Health Education on Smoking in Scotland for the year 1980*

*Signed*    Chairman ___John Brotherston___

Medical Director ___Eileen Crofton___

Courtesy of Loganair

*ASH Award (Action on Smoking and Health)*
*In 1978, the author and his wife, on a Twin Otter flight to Barra, were almost asphyxiated by two passengers smoking throughout the journey and he resolved to have smoking banned on all Loganair flights. In 1980, Loganair received the ASH Award, as the first airline to have a complete smoking ban which was strictly maintained even when the Company began to operate larger aircraft. It was the 1990s before most airlines followed the Company's example.*

British Airways, for example, lodged a successful objection to Loganair's application for Aberdeen–Lerwick on the basis that it would impact on its Aberdeen–Sumburgh route, and less reasonably, objected to Loganair's application for Glasgow–Londonderry on grounds that it would adversely affect traffic on its service to Belfast some eighty miles away.

During this period, Loganair made many applications for route licences, desperate to expand its flying operation and obtain remunerative utilisation for its aircraft. On 2 April 1979, the Company started a double daily Twin Otter service between Glasgow and Londonderry, despite the IRA putting down its marker by blowing up the hangar at Eglinton Airport the day before. Despite The Troubles, which seriously affected Londonderry, the service continued for twenty-eight years without interruption and was brought to an end, not by the threats of the IRA, but by the close attention of Ryanair. Marketing the service often proved difficult. Using 'Londonderry' seemed to alienate part of the local population while 'Derry' the other part. In the early days, Loganair even felt it necessary to have the promotional literature refer to the Glasgow–Eglinton Airport service.

Loganair also started several summer-only services. A weekend service with the Trislander was operated for three seasons from Glasgow to Enniskillen in County Fermanagh using the old wartime airport of St Angelo. Another summer-only service was started in 1980 and operated for four seasons with a Twin Otter on the former Dan-Air and Scottish Airlines route from Prestwick to the Isle of Man. Later, there were also Loganair summer services from Belfast to Blackpool, and Londonderry to Isle of Man and Blackpool.

### End of the British Airways Planning Group
The Planning Group had become increasingly strained, which was hardly surprising in view of Loganair's actions, particularly seeking revocation of British Airways' licence on the Glasgow–Benbecula route. Loganair's announcement that it was acquiring aircraft as big as the Shorts 330 aircraft, of course, was never contemplated when the Planning Group was conceived and this was viewed badly. Loganair undoubtedly acted provocatively during this period in its determination to get access to more lucrative business. British Airways' Scottish management demonstrated great tolerance, but inevitably there were suspicions, especially on the part of British Airways' local staff, and misunderstandings about Loganair, and what it was up to, were common place. Eventually enough was enough. On 18 April 1981, Robert Winyard, British Airways' Manager Scotland wrote that certain actions had been taken at the sole initiative of Loganair and had been viewed badly by British Airways. There would be no more meetings.

During the life of the Planning Group, British Airways had handed over the truly third-level services to Loganair which had been very unprofitable within British Airways' network. Soon after joining the Group in March 1975, the CAA had strenuously reminded the parties that shedding heavily loss-making services clearly made eminent sense for British Airways, but taking on these routes also had to make economic sense for Loganair. Prophetically, the CAA's James Bowley pointed out the pitfalls to Loganair which, he stated, would have to locate aircraft at various places within the Highlands and Islands and seek to find the maximum profitable utilisation of these aircraft from these bases if they were going to achieve a better operational network and better financial result than British Airways had achieved.

Sadly for Loganair, in this, the objectives of the Planning Group were not realised. British Airways' losses from its Scottish operations continued unabated, and Loganair also continued to suffer serious losses even when the subsidy issue was resolved for certain lifeline routes, much to the continuing concern of the Royal Bank of Scotland Board.

### British Airways Highland Division

It was little wonder that in the summer of 1981, following the demise of the Planning Group, there was much speculation about the future of British Airways services in Scotland. British Airways estimated that the losses on its Highlands & Islands services would rise to £5 million in 1982/83. The national carrier had just announced a record pre-tax loss of £141 million and this had prompted Sir John (later Lord) King, its new Chairman, to look at ways of restoring British Airways to profitability as a prelude to 'privatisation' during the life of the Thatcher Government. Jim Harris, head of British Airways' UK and Ireland Division, publicly acknowledged that a British Airways management and trade union team were working on a survival plan and only if they could not find a way to make their eleven Scottish internal routes profitable would they be discontinued. Sir John King commissioned a report from Price Waterhouse and very soon there were 'leaks' that their recommendations would involve British Airways' withdrawal from Scotland.

With all that was going on, in the minds of many this was perfectly feasible. British Airways had announced the ending of Transatlantic services at Prestwick with the termination of its New York and Toronto services at the end of the season, severing a link with Scotland which had begun during World War Two. Its London Heathrow Shuttle services from Glasgow and Edinburgh, which had just lost £1.5 million in the last year mainly due to an air traffic control dispute, now had competition from British Midland Airways with its proposed cut price fares and superior cabin service. The public also saw British Airways facing increased competition with the coming of the four and a half hour Glasgow–London *Advanced Passenger Train* at a time when mounting fuel prices were impacting on air fares. With these severe financial and competitive pressures, British Airways would have to take decisions about its Scottish services.

From Loganair's perspective, British Airways' difficulties presented its own long awaited opportunity. The Loganair Board was even more excited when Sir John King advised the Secretary of State for Scotland that, should the airline pull out of its internal Scottish routes, British Airways would 'co-operate fully with any private sector operator applying to take them over.' This was what Loganair had been waiting for.

Elsewhere, the Western Isles Islands Council, having earlier put the cat among the pigeons with their public soliciting of other airlines to operate the Glasgow–Benbecula service, decided in the end not to support another operator so long as British Airways wished to continue. This effectively killed off Loganair's aspirations in the Western Isles as, under the new CAA Licensing Criteria, for any licence application to be successful would require the support of the Western Isles Islands Council as the 'vitally affected community'.

The author vividly recollects the Shorts 330 demonstration and presentation in Benbecula when real concerns were expressed by local guests about the dimensions of the aircraft's hold and its ability to carry a coffin. He remembers commenting that the Council may have withheld their support for Loganair's application not because of what it could do for the living, but what it could not do for the dead. Not being allowed on to the Glasgow–Benbecula route was a great disappointment for the Company.

Worse still, unlike the Western Isles route, Loganair's application for a licence to operate a new Aberdeen–Lerwick service did have some support in Shetland, but it too was turned down. Never mind, British Airways was surely about to withdraw from its loss-making Scottish routes – Sir John King's comments certainly were interpreted in that way – and there would be many other prizes to be won.

Loganair's eager anticipation of British Airways' withdrawal from Scotland, meanwhile, was being shared by other airlines. Dan-Air with its HS 748s, Alidair with its Viscounts and Air Ecosse with its Bandeirantes were all circling round, ready to pounce. Alas, all hopes were to be dashed. At a Licence Hearing in Aberdeen in September 1981, chaired by Raymond Colegate, Hugh Welburn outlined British Airways' survival plan which had been agreed by both management and unions. The union's involvement was crucial.

The British Airways Trades Union Council comprised engineers, ground staff and cabin staff as well as pilots and they were involved in a remarkably selfless exercise. A new 'Highland Division' would be formed. Forty-four seat HS 748 aircraft would replace seventy-two seat Viscounts which would improve the viability of the Scottish internal services. The staff complement would reduce from 604 to 184. New working practices were to be adopted. The pilots, in particular, were prepared to accept new terms and conditions, including, crucially, multi-tasking. Only the Viscount cabin staff on the Scottish routes would make no concessions and none would agree to join the new Highland Division. For its part, British Airways management, including Sir John King, and Colin Marshall, British Airways' Chief Executive, welcomed a plan which would allow them to continue in Scotland and avoid the wrath of the powerful Scottish political lobby.

The CAA welcomed these initiatives and effectively announced a moratorium on route licence applications for British Airways' Scottish routes for a period of two years, thereby giving Highland Division every chance to achieve its objectives:

> If British Airways fails in its present plans so that the promised gains in efficiency and benefit to users are not achieved, there is argument for licensing competing or substitute services, perhaps more strongly than before. The better outcome is that British Airways should not fail in its plans, not least because they set a pattern for adoption elsewhere.

Photo: Iain Hutchison

*Known affectionately as the 'Budgie', the Hawker Siddeley 748 was acquired by British Airways in 1975 and used initially on the Aberdeen–Sumburgh route. The HS 748 was being used increasingly in North Sea oil support and Dan-Air had fourteen aircraft and thirty-six crews based in Aberdeen by the later 1970s.*

*As part of the British Airways survival plan when Highland Division was formed in 1982, three HS 748s were leased from Dan-Air to replace the remaining Viscount fleet. By 1984 British Airways was introducing the more expensive, but quieter and more fuel-efficient, British Aerospace Super 748 on the Highlands and Islands routes.*

## Down and Out?

Historic, ground breaking, revolutionary, it may have been, but the creation of British Airways Highland Division was a serious setback for John Burke and the Loganair management. As no new Scottish route licences would now be won from British Airways for the foreseeable future, the Loganair Board felt they had no choice but to mount a greater challenge to British Airways where it could. That could only be on those routes which Loganair already shared with British Airways, namely Glasgow–Inverness and Glasgow–Belfast, by increasing frequency on both routes from January 1982. However, having been given such a reprieve after their near total withdrawal from Scotland, British Airways Highland Division was now determined to fight its corner on these two routes.

On the Inverness route, for example, it lowered its fares and moved its flight times to within a few minutes of Loganair's timetable, and paid for an expensive full-page, rather negative advertisement in the *Glasgow Herald*, *Scotsman* and *Press & Journal* newspapers. The advertisement depicted the British Airways aircraft flying above the clouds and, at a much lower level, bumping through dark cumulus were aircraft clearly recognisable as Loganair's Shorts 330, Bandeirante, Twin Otter and Islander. This was a source of great irritation to Chairman John Burke and to the author who questioned whether this was a legitimate use of taxpayers' money by the state airline.

On the charter front, Loganair's situation was also depressing. The Chevron contract, which had involved seven Twin Otters, had now been terminated. Prospects in Aberdeen for fixed-wing, oil charter work had been seriously affected by the advent of a new generation of longer range helicopters. The Puma and the Chinook were now on the scene and were capable of flying direct from Aberdeen to the oil platforms in the East Shetland Basin, obviating the need for fixed wing aircraft. The prevailing low oil price was now also discouraging oil exploration activity and fixed-wing support had dwindled. Such fixed-wing work as remained was yielding increasingly modest margins in the diminishing but still very competitive market. The emphasis was likely to be on the faster, longer range aircraft to fly to Ireland, the South West approaches and the near Continent, and this meant that Bandeirantes and Jetstreams would be more acceptable than the Twin Otter.

The Loganair Board's action plan was to scale back its charter base in Aberdeen by reducing the charter fleet to two Bandeirantes and one Twin Otter, and selling the Aberdeen hangar and office complex. There was no lack of effort and some Twin Otter work was gratefully obtained abroad, for example from Schreiner Airways for Red Cross operations in Angola, and oil-related work in Sudan. Twin Otters and the remaining three Trislanders were to be sold as soon as possible.

In accordance with its own survival plan, Loganair stepped up its direct competition with the new Highland Division on the Glasgow–Belfast (Aldergrove) route, and the Glasgow–Inverness route with additional services at peak times. British Airways immediately retaliated with tariff reductions and improved timetables and Loganair struggled, particularly on the Belfast route.

## *Fighting back - with some success*

Loganair had been a customer of Short Brothers since 1969 when the Company had bought the Skyvan and, more recently, the two Shorts 330 aircraft. Short Brothers had its own airfield at Sydenham and Loganair now approached the Shorts Board with a view to using its airport for scheduled service purposes. Eventually, Shorts agreed that operators of its Shorts equipment could use Sydenham, to be known as Belfast Harbour Airport (later, of course, Belfast City). Loganair switched its Glasgow service from Aldergrove and on 7 February 1983 operated the first scheduled service at Belfast Harbour. Passenger traffic improved quickly by some forty percent, because of the proximity of the airport to the city centre, compared to the seventeen-miles-distant Aldergrove and. Loganair's other services quickly followed suit. Belfast Harbour became an increasingly important Loganair base.

With no prospect of being licensed to challenge British Airways on its Scottish internal services, Loganair looked cross-border at British Airways' Edinburgh–Manchester service which had been receiving much passenger criticism. The service was linked to British Airways' Manchester-Düsseldorf route and was prone to severe delays which knocked on to the Edinburgh sectors. British Airways' poor performance had caused passenger traffic to fall from 44,000 to 25,000 in the last four years. Loganair saw its chance and decided to apply for the route and seek revocation of British Airways' licence. Despite route licence applications from Air Ecosse and Alidair, Loganair was awarded the licence after British Airways, at the last minute, announced it was axing seventeen routes, including Edinburgh–Manchester, as it strived to improve its overall financial performance. This was a huge breakthrough and suddenly gave Loganair a scheduled service with a passenger volume it had been seeking for years. The service commenced in October 1982 with a Shorts 330 three times daily. This was better, much better.

## *Bank Concerns*

Meanwhile, time was running out. The Royal Bank of Scotland Board was viewing Loganair's mounting losses with concern. John Burke had given the Company's management every opportunity to expand and trade its way out of its loss-making operations. He had accepted each business case for acquiring Twin Otter aircraft followed by Bandeirantes and Shorts 330s. The Company had lost £700,000 in

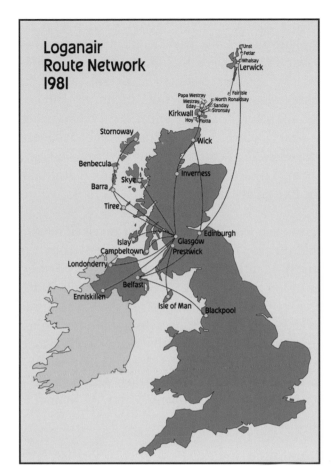

Loganair
Route Network
1981

Unst
Fetlar
Whalsay
Lerwick

Fair Isle
Papa Westray
Westray    North Ronaldsay
Eday    Sanday
Kirkwall    Stronsay
Hoy    Flotta

Stornoway

Wick

Benbecula

Skye    Inverness

Barra

Tiree

Edinburgh
Islay    Glasgow
Campbeltown    Prestwick
Londonderry

Belfast
Enniskillen

Isle of Man
Blackpool

*Through the auspices of the British Airways/Loganair Joint Planning Group, services to Campbeltown, Islay, Tiree and Barra had been transferred from British Airways, as had the Stornoway-Benbecula and Inverness–Wick–Kirkwall service. The Inverness–Edinburgh service had become an important route for Loganair.*

*By 1981, the route map has expanded greatly. Northern Ireland has become an important new market with Belfast being served from Glasgow and Edinburgh, while Londonderry and Enniskillen are also served from Glasgow. The first service to England is also operated, but not from Scotland. The service to Blackpool is operated from Belfast. Inverness is now served from Glasgow.*

1979/80 and £500,000 the next year. Prospects for 1981/82 looked even worse. Of the eagerly sought after routes transferred earlier from British Airways through the auspices of the Planning Group, only those in receipt of subsidy were actually profitable. Of the others, Glasgow–Campbeltown was particularly disappointing, and Glasgow–Islay only a little less so. On the oil charter front, the premature termination of the Chevron contract, for which the Company had expressly purchased six Twin Otters, was extremely damaging. John Burke's readiness to allow the Company to equip itself to be in position to benefit from British Airways' widely-assumed withdrawal from their Highlands and Islands routes had been cruelly misplaced. The Company's action plan, involving essentially the disposal of Trislanders and Twin Otters, would take some time to implement. Loganair's financial position was dire and the Royal Bank Board was looking for a solution.

There had been a number of concerns also within the Bank about the management of Loganair. There had been some disquiet, for example, about extensive press coverage of Duncan McIntosh's relationship with Lady Strathcona, whom he later married. More particularly, there was a strongly held view that this had been a serious distraction for Captain McIntosh from his Loganair business at a time when the Company's extreme financial difficulties demanded his full attention. Duncan McIntosh took early retirement from the Company on 31 December 1982 having been involved since Day One some twenty years before. Gilbert Fraser, who had been Engineering Director since 1975, now assumed the position of Engineering Manager with overall engineering responsibility being taken over by an engineering consultant, Roy Bonney. Scott Grier, who had been Financial Director since 1977, took over as Managing Director on 1 January 1983.

The Royal Bank's own situation had also changed. There had been a quashed merger with the Hong Kong and Shanghai Bank, quickly followed by a takeover bid by the Standard Chartered Bank. The Government would step in with its so-called 'Tartan' defence and vetoed any foreign takeover of a Scottish icon like the Royal Bank. In the process, however, the anomalous situation of the Bank owning a regional airline was highlighted. Owning a regional airline which was also making substantial losses must have appeared to most observers as quite bizarre. The Royal Bank Board clearly had to address the Loganair issue. Its clear message, delivered by John Burke to new Managing Director, Scott Grier, was to eliminate the £1 million loss and achieve a break-even situation to clear the way for the sale of the Company within the year.

### Finding an honourable exit

Clearly under pressure, in April 1983, John Burke suggested closer collaboration with Air Ecosse. It was a competitor but its financial position in recent years had been stronger than that of Loganair – a profit of £465,000 had been reported the previous year – but it was clear that Air Ecosse was now also finding business difficult. Three of its fleet of ten Bandeirantes had just been sold. The fundamental flaw in its business strategy was the serious imbalance between night mail operations for Datapost, which required many aircraft but only for a few hours flying every night, and its very marginal daytime services. Air Ecosse too was withdrawing from the ad hoc charter market and with the sale of the surplus aircraft there would be redundancies among the one hundred and sixty staff. Recently, Loganair had been subcontracted by Air Ecosse to fly several of its night operations which had allowed it to reduce its aircraft fleet. Clearly collaboration between the two airlines had been mutually beneficial. John Burke wanted to know if this collaboration could be extended.

Initially, there was no thought of a merger or indeed the sale of Loganair to Air Ecosse. However, in the following months many meetings took place with Henry

Cameron, Chairman of Air Ecosse, who was being advised by Ewan Brown of Noble Grossart. Ewan Brown had carried out a feasibility study which indicated that a merged business could be profitable and a formal offer for Loganair was made to the Board of the Royal Bank. For a variety of reasons – not the least of these, the author would like to believe, were his sustained and very robust counter-arguments to any takeover by Air Ecosse – and worries about Air Ecosse's middle and longer term problems, the Bank decided to reject the offer.

The Loganair Board felt one last-ditch approach should be made to British Airways in the belief that the new Highland Division was doomed to failure. These Scottish services, had, after all, incurred serious losses every year since 1947 and British Airways was busy getting its house in order for privatisation. Loganair argued that a merged company would be a good mix of appropriately-sized aircraft and the aircraft fleets of both Companies could be reduced in number. Moreover, staff savings could be realised at several airports where both Loganair and British Airways had ground staff.

John Burke and the author met Gordon Dunlop, British Airways' Finance Director, and David Hyde, by now Head of British Airways' European Division, but it quickly became clear that British Airways would entertain no proposals until the work practices established by the new Highland Division had been implemented elsewhere in the British Airways operation. British Airways' old dilemma, to receive or not to receive subsidies for the Scottish internal services, also cropped up again. British Airways clearly was still not prepared to accept the conditions of transparency that were attached to the granting of Government subsidies. In any event, the Royal Bank Board was by now showing no enthusiasm for any venture that would involve the Bank in any greater commitment in aviation. They wanted out.

Such was the determination to find a solution to its Loganair problem, that the Bank's Chairman, Sir Michael Herries, entered the fray and approached Peter Buckley of British and Commonwealth Holdings to explore the possibility of a tie-up between its subsidiary, Air UK, and Loganair. A meeting between the author and Philip Chapman quickly confirmed that Air UK had no interest whatever in a joint venture with Loganair. A similar approach by John Burke to the Scottish Transport Group, through its Chairman, Bill Stevenson, was no more fruitful.

Following a presentation by the author to the Board of the Royal Bank, John Burke requested a deferment of any decision for six months in view of Loganair's much improved financial results and outlook. In less than a year, the fleet had been reduced to fifteen aircraft by the disposal of all six Trislanders and three Twin Otters. A better mix of business with this reduced fleet was now allowing the Company to achieve an annual turnover similar to what had previously been earned by twenty-four aircraft.

The new Edinburgh–Manchester scheduled service was doing well. A breakeven result for the Company for the current year was certainly possible. Alas, too late. The Bank Board was unmoved and reaffirmed disengagement.

It was becoming frenetic. With a distinct feeling of *déjà vu,* the Board then considered an approach from Peter M Kaye of Channel Express, but turned down his overtures, just as John Burke had first turned him down some eleven years earlier. Financial information about Loganair was requested and provided to Harold Bamberg, who had headed up British Eagle International Airlines more than a decade earlier, but this also came to nothing.

### British Midland Airways

The only credible approach came from Michael Bishop of British Midland Airways. He had recently formed Manx Airlines with BMA taking seventy-five percent of the equity, and Air UK owning twenty-five percent. He now proposed a similar arrangement, with the Royal Bank retaining twenty-five percent of Loganair, but any continuing interest by the Bank was quickly discounted. Terms were negotiated on a Net Asset Value basis in view of the Company's track record of non-profitability. The timing of the sale transaction was influenced by the significant Group Relief arising from Loganair's losses which were available to the Royal Bank. The deal therefore had to be delayed until after 30 September 1983, the Royal Bank's Balance Sheet date.

In the event, the transaction was delayed further and scheduled for 3 December. Tragically, John Burke was killed in an accident on Meall Buidhe in Argyll while walking with his friend, Dr James Manson, only ten days before the sale of Loganair had been completed. He had striven throughout the entire period he was Chairman of Loganair to achieve an honourable sale of the Company. He was always mindful of his responsibilities to the Bank Board, but also always acutely aware of the importance of Loganair's role and the contribution it was making to the economic and social welfare of the Scottish Highlands and Islands. He had appreciated the enormous efforts made by Loganair management and staff to reach for the Holy Grail of profitability and remained supportive to the end. Loganair was saddened by the cruel twist of fate which prevented his being there to complete the deal. Charles M Winter of the Royal Bank was appointed Chairman of Loganair for the short period until 3 December 1983 when British Midland Airways acquired the Company and Scott Grier acquired twenty-five percent of the equity.

### The end of an era

Two long serving stalwarts left the Company at this time. Captain Ken Foster had joined Duncan McIntosh from the RAF in 1963, having flown Meteors and Javelins at Leuchars. He progressed with Loganair from pilot, to Chief Training Captain, to Chief Pilot and, in 1975 was appointed Director of Flight Operations. He played a vital role, not only in developing the new routes, but was also personally involved in establishing new airstrips to serve remote communities - especially for use by the Scottish Air Ambulance Service for which he was directly responsible. Captain Foster was awarded the Queen's Commendation for Valuable Services in the Air by the Lord Lieutenant of Renfrewshire, Major J D M Crichton Maitland, in 1981.

Also ending a long association with the Company at this time was Jim Harter. He had been appointed to the Loganair Board by the Bank at the time of its takeover of Loganair in 1968 when his firm became the Company's auditors. In 1971 he had to step down from the Board but continued as financial adviser, first to Hamish Robertson and then to John Burke. Both Chairmen relied heavily on Jim Harter's judgment and independent financial advice. In the earlier Bank years, he was responsible for preparing the many tenders for charter work or documentation in support of the numerous applications that were being made to the CAA for route licences and tariff increases. Importantly, when the Bank took over, he established formal systems in the Company where previously there had been none. He was a valuable sounding board, not only for the Chairman, but for all the members of the Board.

After having left Loganair on 31 December 1982, Duncan McIntosh was determined to stay in aviation and formed a new company, Duncan McIntosh Aviation Limited, with plans for a second-hand Britten-Norman Islander charter operation. Shortly thereafter, it amalgamated with several other small aviation companies into the Air Charter Scotland Group chaired by Sir Hugh Fraser. Later, Captain Mac became involved in Malinair which flew services between Glasgow and Donegal. None of these enterprises enjoyed great success. Captain Mac was also suffering from increasingly poor health and it was a sad irony that he died on Loganair's 25th anniversary. In his obituary in the *West Highland Free Press* to 'Duncan McIntosh "Friend to Island Communities"', Brian Wilson wrote:

> The growth of air communications over the past couple of decades has been a vital asset of island development in general. Nowadays, it is all taken very much for granted. The fact that the network exists, and works so well, is due in no small part to the fact that Duncan McIntosh was not only involved, but also cared.

Back in 1972, when he became Loganair Chairman, John Burke was interviewed by the Press about his taking his Private Pilot's Licence. What did the Bank have to say about his new hobby? 'They have not expressed an opinion about flight safety. I am at no greater risk than driving my car or climbing a hill.'

*The all male management influence is evident in the uniform of around 1970.*
(Courtesy of Loganair)

*In the early 1970s, before the employment of cabin attendants, red and white check was favoured for passenger services staff.* (Photo: Randak Design)

*Uniformed representatives of airlines serving Scotland in 1979 pose at Glasgow Airport. From left, Icelandair, Dan-Air, Loganair, KLM, Pan Am, Japan Air Lines, Britannia Airways, British Midland Airways, Aer Lingus, British Caledonian and British Airways. From a bygone age!*
(Courtesy of Loganair)

*The red and black uniform introduced in 1977 and worn by all female staff members of the Company.*
(Photo: Randak Design)

*The new Thomson tartan uniform introduced in 1984 by Debbie McIntyre, Irene Graham, Carolyn O'Neill, Elizabeth Taylor, Pat Gordon and Claire Fernie in front of the Embraer Bandeirante.* (Photo: Bob Nicholson)

*Concorde was conveniently on hand to mark the start of the Franchise Partnership with British Airways in 1994. The Company cabin staff now wears the British Airways uniform, replacing the Loganair grey and tartan.* (Courtesy of Loganair)

*Following the termination of the British Airways Franchise Agreement, the Company became a Franchise Partner of Flybe, the largest regional airline in the UK, with effect from October 2008. Here Loganair's Dundee cabin staff, Jo Donlon, Jillian Robertson and Laura Burnett are wearing the new uniform.* (Courtesy of Loganair)

# Chapter 3

# The British Midland Years
# 1983–1997

## *British Midland Airways*

Michael Bishop, Chairman of British Midland Airways, apparently had the choice of three other independent airlines while negotiating with the Royal Bank to buy Loganair. It was Loganair in Scotland, however, that presented the best fit for Michael Bishop's plans. Since Prime Minister Margaret Thatcher's Conservative Government had begun to open Britain's skies to a little more competition, BMA had won the licence to compete with British Airways on the Glasgow–London Heathrow route in October 1982. He had done well. Undaunted by his previous failure to win a licence to operate on the London Heathrow–Belfast route, Michael Bishop had applied for the Heathrow routes to both Glasgow and Edinburgh. The Licence Hearings were bitterly contested by British Airways and, despite BMA having strong support from Scotland, his application was turned down by the CAA. Showing his characteristic determination, he appealed to the Secretary of State for Trade and Industry. Press reports of personal intervention in the matter by the Prime Minister herself remain purely speculative, but what is sure, the CAA ruling was overturned despite the fact that the Government was preparing British Airways for privatisation, and competition on its prized domestic trunk routes was the last thing British Airways needed at that moment.

British Midland Airways had been operating under that name since 1964, having begun as Derby Airways in 1947 and originally founded as Air Schools Limited in 1938. As Derby Airways, and more recently as BMA, it had operated scheduled services for some thirty years on the 'Rolls-Royce' route from East Midlands to Glasgow and was therefore no newcomer to Scotland.

Acquiring Loganair was part of a wider strategy. Michael Bishop was setting up a regional network for BMA which included Liverpool, Teesside, Leeds and the Isle of Man. BMA, along with Air UK, had recently formed Manx Airlines, BMA taking the controlling seventy-five percent stake. He explained this strategy was based on the post-Jimmy Carter Deregulated US model, citing US Air's Allegheny Commuter 'although on a different scale'. To the BMA network would now be added Loganair's twenty-six domestic destinations.

In winning the Heathrow–Glasgow route licence, BMA had broken the mould, or more accurately, had broken the long-established British Airways monopoly. BMA now offered a superior in-flight service which contrasted with British Airways' no frills Shuttle service on its Trident aircraft. The Glasgow-Heathrow service was followed early in 1983 by a similar service from Edinburgh. Michael Bishop was well aware that British Airways still had the competitive edge as its Highland Division services offered connections to its London services. Loganair also had a number of air services from different airports in the Western and Northern Isles to Glasgow and Edinburgh, and it would now 'feed' the BMA services.

Michael Bishop was not content merely to take on British Airways on the UK trunk routes. He had applied also to commence services to New York from Manchester and Glasgow. Naming Glasgow was controversial. Michael Bishop wanted Glasgow because of its connecting services, and not Prestwick which, in accordance with Government's Scottish Lowland Airport Policy, was Scotland's designated transatlantic 'Gateway'. Michael Bishop believed that Loganair's services linking Glasgow with many parts of Scotland would be an important factor in presenting BMA's case for the transatlantic licence application. BAA, the owners of both Prestwick and Glasgow airports, objected to the BMA application.

BMA had plenty of support in Scotland for its proposed Glasgow–New York service, none of which was more vociferous than that from Dr Michael Kelly, Lord Provost of Glasgow, who had been concerned at the lack of direct services from Scotland to the US since Laker Airways' collapse in 1982 and indeed British Airways' own withdrawal from Prestwick in 1983. The CAA was also persuaded by the BMA case and awarded it a licence for Glasgow–New York, stating that the service would have a better chance of success from BMA's existing hub at Glasgow and nearer to the major population catchment area. BAA's Chairman, Norman Payne, lodged an appeal with Secretary of State for Transport, Nicholas Ridley, arguing that Prestwick was 'far from dead'. Nicholas Ridley overturned the CAA decision and then called for a review of Scottish Airport Policy. George Younger, Secretary of State for Scotland and MP for Ayr (and later Lord Younger of Prestwick), in particular fought a rearguard action to retain Prestwick's transatlantic status despite it being cut off from Scottish mainstream aviation. It was 1990 before Glasgow Airport had its transatlantic services.

### The Loganair Board
The new Loganair Board comprised Michael Bishop as Chairman, Scott Grier continuing as Managing Director, and members of the BMA Board – John T Wolfe and Stuart M Balmforth, both 24.5 percent shareholders and General Manager and Company Secretary respectively of BMA. The Board was completed by Grahame N Elliott, a Manchester Chartered Accountant and friend and adviser to Michael Bishop.

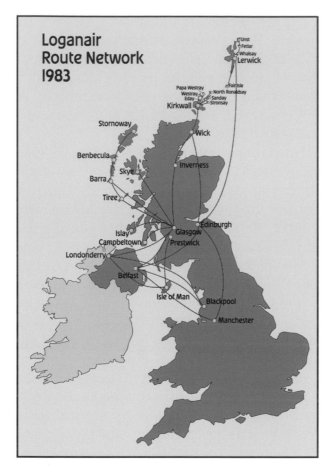

*By the time British Midland Airways acquired Loganair in December 1983, the Company's route network had expanded strongly. It was competing with British Airways on the Glasgow–Inverness route, and the non-stop Twin Otter service between Edinburgh and Lerwick was now going from strength to strength. Northern Ireland had become an important market with Belfast being linked with Glasgow and Edinburgh and Manchester, while Londonderry now had services to Glasgow, Blackpool and the Isle of Man. Most excitingly, Manchester was now an important base with services to Edinburgh and Belfast. Manchester–Glasgow would follow shortly.*

Austin Reid was appointed to the Board when he joined British Midland as Finance Director in 1985. All these Directors remained on Loganair's Board throughout the years of British Midland ownership. The author remained the only Loganair and Scottish representative. Michael Bishop firmly turned down a request in 1990 by the Industry Department for Scotland for a Non Executive appointment on the Loganair Board. Similarly, a later request by the author to promote Loganair's Chief Executive–Technical, Trevor Bush, to the Board was refused. In this respect, the composition of the Loganair Board differed from that of British Midland's other subsidiary, Manx Airlines, where local Isle of Man interests were represented.

### Fleet Development
The Loganair that BMA bought in December 1983 was by then well out of the financial intensive care that it had been under for many years. Significant progress had

been made in the last year. Some 210,000 passengers were being carried annually on fifteen aircraft, achieving a turnover of £10 million (BMA annual turnover was £58 million), and 220 staff were employed. A near breakeven-position recovery from the previous year's disastrous £1 million loss had been possible because there was now a better business mix, but also because of the successful reduction in aircraft in the fleet. In December 1983 there were six Britten-Norman Islanders for the inter-island services in Orkney and Shetland and the Scottish Air Ambulance Contract, five de Havilland Twin Otters employed on both scheduled and charter services, and two Embraer Bandeirantes used primarily for oil charters and mail contracts. One of the two Shorts 330 aircraft had been replaced a few months earlier by the bigger, 36-seat Shorts 360. The fleet reorganisation had just been completed with a Fokker Friendship leased from BMA with heavy checks outsourced to Air UK and line maintenance by Dan-Air at Manchester.

Photo: Iain Hutchison

*The Company leased the first of its three 44-seat Fokker Friendships in November 1983, a month before the takeover by British Midland. The aircraft became the workhorses on the Manchester routes to Edinburgh, Glasgow and Belfast before being replaced by the 64-seat BAe ATP aircraft in 1989.*

BMA already had a Shorts 330 operating its East Midlands–London Heathrow service. Manx also had a Shorts 360. Michael Bishop was an enthusiastic supporter of Short Brothers aircraft. He was not alone, and Alex Roberts, Shorts Sales and Marketing Director, was having great success. While the largest Shorts fleets were with US commuter airlines such as Mississippi Valley Airlines, Allegheny Commuter, Simmons Airlines and Command Airways, there were also many European customers. Apart from the BMA group of airlines, Air Ecosse in Scotland, Inter City/Guernsey Airlines, Genair, Air UK and, in Ireland, Avair and Aer Lingus, were all Shorts operators. Michael Bishop asserted, at a time of high fuel costs, that the fuel consumption of the Shorts, and indeed their operating costs generally, compared most

*While still in the ownership of the Royal Bank of Scotland, the Company took delivery of its first thirty-six seat Shorts 360 in March 1983. With its high, spacious cabin and improved passenger comfort, the Company was able to offer a superior inflight service in competition with British Airways and, more particularly, British Airways Highland Division. The Shorts 360 was a familiar sight on all Loganair's main routes, particularly in Scotland, and was withdrawn from service only in 2005. By the 1990s, however, it had lost much of its earlier passenger appeal as it was unpressurised and compared unfavourably with British Airways' HS 748 and BAe ATP.*

favourably with the older turbo-props in the BMA fleet and he regularly substituted the Shorts for the Viscount, Herald and Fokker Friendship.

He quickly decided to trade in Loganair's remaining Shorts 330 for a second Shorts 360 and stated the Company's intention to increase the Shorts 360 fleet, not only in Loganair, but in Manx and BMA also. He reaffirmed the longer term intention to dispose of the two Bandeirantes even although it would have serious implications for Loganair's struggling Aberdeen oil charter base and for its night mail contracts. The number of Twin Otters was also to be reduced. While suitable for many of Loganair's Highlands and Islands scheduled services, oil charter work for the Twin Otter was diminishing. Importantly also, Twin Otters attracted a good resale price and cash flow considerations were becoming increasingly important.

Loganair was operating five Shorts 360s by 1986 when Michael Bishop announced a £9 million group order for three new Shorts 360 aircraft, two for BMA for its links between Heathrow and East Midlands and Birmingham, and the third for Loganair. At the same time, British Midland transferred two Shorts 360 aircraft to Manx to allow it to take over the Heathrow–Liverpool service from British Midland. Michael Bishop remained enthusiastic: 'There is no doubt that many low volume routes which are persistent loss makers have been turned into profit by the operation of the Shorts 360.' Praise indeed for the 'Flying Shed' as it was often irreverently nicknamed.

The Company's main Scottish services were operated by a combination of Shorts 360 and Twin Otter operations and this was proving to be a successful formula. It was a great shock to everyone in the Company, therefore, when on 12 June 1986, Loganair suffered its first ever fatal accident. A Twin Otter, flying the afternoon Glasgow–Islay scheduled service, crashed on a hillside near Lagavulin on Islay. The co-pilot, First Officer Christopher Brooke, was killed, and Captain David Ilsley and the fourteen passengers were injured. Following the accident, Michael Bishop and John Wolfe had the Company's operations management reorganised with two British Midland managers seconded for a period. The ensuing Air Accident Investigation Branch (AAIB) enquiry found the commanding pilot culpable. The AAIB recommendation that all Twin Otters have a wing restraint fitted was accepted by the Company and implemented.

In late 1986, Loganair was called upon to help out its newly formed sister company, Eurocity Express, which was planning to commence de Havilland Dash 7 services from London City Airport from October 1987. Two Dash 7s were taken on short-term leases pending delivery later in 1987 of two new Dash 7 aircraft from de Havilland. Loganair's Engineering Manager, Gilbert Fraser, surveyed and accepted the Newmans Air aircraft in Christchurch, New Zealand, and the aircraft were taken on to the Company's Air Operator Certificate (AOC).

Loganair was given the responsibility of operating one of the aircraft from April 1987, using the time for proving and training purposes, until going on to Eurocity Express service at the London Docklands STOLport. Meanwhile, for Loganair to find a suitable task for the Dash 7 within its existing network was not easy. It was not suited to the Highlands and Islands services, the Company's efforts to interest Shell and Chevron in the aircraft for oil support work were unsuccessful, and in the end the Dash 7 was used on Loganair's Manchester– Belfast route. The second Dash 7 was leased to British Midland.

Photo: David Dyer Collection

*The 50-seat de Havilland Dash 7 in the 'business suit' livery of Loganair's sister company, Eurocity Express. Loganair operated this aircraft on its Belfast–Manchester route from April 1987 while Eurocity Express awaited approval for operations at London Docklands STOLport. The company name was later changed to London City Airways.*

*And then something new – the BAe ATP*

At the Paris Air Show in 1985, Michael Bishop and British Aerospace Chairman, Sir Raymond Lygo, signed a Purchase Agreement for British Midland for three firm orders and two options for the new BAe ATP (Advanced Turbo Prop) aircraft. British Midland would be the launch customer for British Aerospace's sixty-four to sixty-eight-seat successor to its forty-four-seat HS 748 and it was the first time that British Midland had ordered new aircraft. In early 1985, it was not just Loganair which had cash flow pressures. British Midland also was stretched and an important part of the arrangement was for British Aerospace to buy British Midland's seven ex-South African Airways Viscounts, several of which had already been withdrawn from service. Doing the deal was clearly very important to British Midland. In *Diamond Flight: the story of British Midland*, Bill Gunston wrote that it was a decision taken by its three principal shareholders 'after careful, but not over-prolonged evaluation' and the rest of the airline was informed later. It was a decision that would have implications for Loganair. The Group management duly assembled at Woodford, Manchester, on 6 August 1986 to watch British Aerospace's test pilot, Robbie Robinson, as he successfully undertook the first flight of the BAe ATP.

British Aerospace had high hopes for its new aircraft. Sales of its world-beater at £7 million each, BAe believed, could reach £2.5 billion and keep the 6,000 BAe staff at Chadderton and Woodford busy for the next twenty years. Charles Masefield, BAe Divisional Director, admitted that in the past, British planes like the Comet, the Viscount and the VC10 were too late and too costly. The ATP would be different. Alas, the ATP programme was hugely problematical.

British Midland had been expanding its scheduled service network since it had made its great breakthrough when it was awarded the licences to operate on the Glasgow, Edinburgh and Belfast to London Heathrow routes. It had become a major operator at Heathrow and in 1986 had undergone a total rebranding with the introduction of its *Diamond Service*. A major further expansion was planned when no less than eleven international route licence applications were lodged. This, of course, required significant investment in many more jet aircraft. Additional finance would also be needed following the BMA Board's decision to simultaneously replace the turbo-prop fleet of Shorts 360s and Fokker Friendships with the new BAe ATP aircraft. The sheer scale of this aircraft investment programme contrasted with Michael Bishop's first 'fleet' purchase of seven Viscounts from South African Airways for £95,000. Clearly it was time to reorganise the Group structure.

Since 1983, British Midland Airways had held seventy-five percent of the equity of Manx Airlines and Loganair. A new Holding Company, Airlines of Britain Holdings Limited (ABH), was formed in 1987 and now had three wholly owned subsidiaries,

*ATP signing - A momentous occasion for the British Midland Group Companies. During the Paris Air Show in 1985, orders were placed initially for three BAe ATP aircraft and options for two more. Sir Raymond Lygo presents Michael Bishop, Chairman of British Midland, with a model of the ATP in British Midland colours after the initial contract signing. Left to right: John Wolfe, Director British Midland, Terry Liddiard, MD Manx Airlines, Sir Raymond Lygo, Chairman of British Aerospace, Scott Grier, MD Loganair, Michael Bishop, Chairman of British Midland, Charles Masefield, Sales Director British Aerospace and Norman Barbour, Director British Aerospace.*

Photo: British Aerospace

British Midland, Manx and Loganair with the twenty-five percent minority holdings held by Air UK in Manx, and held by Scott Grier in Loganair, being bought out. Scandinavian Airline Systems (SAS), having recently failed to acquire British Caledonian Airways (BCAL) and now with a real interest in Heathrow and its interline feed, paid £25 million for a 24.9 percent shareholding in ABH which would, of course, be important in underpinning the Group's expansion. Through this development, SAS found itself having a 24.9 percent holding in Loganair. Any historical concerns that Loganair's Orkney and Shetland customers may have had about this latest Viking raid proved to be groundless. SAS seemed to take little or no interest in what Loganair was up to.

### Policy Review and British Airways largesse
In December 1983, Nicholas Ridley, the Secretary of State for Transport, asked the CAA to advise him on the implications of privatising British Airways 'for competition and the sound development of the British airline industry.' The resulting 1985 Policy Review offered no immediate benefits to Loganair, not least of all because British Airways Highland Division was set to continue. Although the White Paper recognised the special problems of Scotland's air services, the CAA stated that it was unable to finalise its proposals for 'the exceptional treatment of vulnerable routes to outlying communities.' This was a huge disappointment to the Loganair Board. An arrangement, however, had been made with the Government for British Airways to make a total of £7 million available in £450,000 tranches to airlines for the development of up to fifteen international routes from provincial airports. Not surprisingly, the prospect of sharing this largesse was of great interest to the Company, not least of all because of Loganair's considerable debt burden at a time when bank

interest rates had moved from nine percent to fourteen percent. By now, reducing the Company's bank borrowings had become a priority.

The Board was already looking at ways to restructure the Company financially, including approaching the Scottish Institutions and Trust Funds. There was dialogue with the funding organisation ICFC (Industrial and Commercial Finance Corporation), but it was already heavily committed to the aviation sector through its investment in British Caledonian. Nor had anything come of discussions the Chairman had with Schroder Wagg about a possible placing on the Stock Exchange. To help cash flow, there had been the thought that the Twin Otters could be sold and leased back. Now there emerged the possibility that the Twin Otters could in fact be sold and replaced by Dornier 228-200 aircraft. This aircraft type, unlike the slower Twin Otter, could perhaps operate some non-stop international services and qualify for the British Airways largesse, as well as undertake Loganair's current Twin Otter tasks. Indeed, it could operate better than the Twin Otter in the case of the long Edinburgh–Lerwick service where the Dornier would cut thirty-five minutes off the scheduled Twin Otter flight time.

Accordingly, Loganair lodged six international route licence applications before the closing date of 12 April 1985, for services to commence by 30 March 1986. These were Glasgow to Brussels and Cologne, Edinburgh to Copenhagen and Brussels, and Manchester to Rotterdam and Cologne. Loganair was confident that the CAA, which had been given responsibility for allocating the funds for services to all six provincial airports, Manchester, Birmingham, Newcastle, Glasgow, Edinburgh and Aberdeen, would find its own application for Glasgow and Edinburgh of superior quality to the other applications for these airports.

The total cost of acquisition of four Dorniers would be £7-7.5 million. The likely sale price of the Twin Otters after mortgage repayments would allow a most welcome cash benefit of around £400,000 per aircraft. Moreover, tax-based leases for the Dorniers were available from several finance houses, but the most intriguing offer was the package from Peter M Kaye, who had previously shown interest in acquiring Loganair back in 1972 and 1983. He would not only finance the new Dorniers, he would take over the four Twin Otters plus spares and the whole arrangement would produce a £1.5 million cash injection to Loganair. However, it was all to no avail. After discussion, and perhaps because of the image of the unpressurised Dornier flying across the North Sea in bad weather, the Board decided that Loganair's policy should be to identify established routes using bigger aircraft.

Again, with the British Airways largesse still in its sights, the Company did look briefly at the Jetstream 31 manufactured at nearby Prestwick for these European routes, but without STOL capability the Jetstream 31 was unsuited to several of the

Company's Highlands and Islands routes. The Dornier project was postponed indefinitely. Meanwhile Peter Kaye acquired one of the Twin Otters. Of the eighty-two or so applications from eighteen airlines for a share of the British Airways largesse, most fell away by the time of the Hearings. Birmingham Executive, with its Jetstream 31 fleet, was most successful, being awarded four routes, Birmingham to Amsterdam, Dusseldorf, Frankfurt and Stuttgart, and subject to traffic rights also to Oslo. Birmingham Executive received just under £2 million with a helpful upfront £900,000 payment. Dan-Air also benefited and was awarded three routes: Manchester to Amsterdam, Oslo and Stockholm. Nearer home, Ace Aviation, later Chieftain Airways, won routes from Glasgow to Brussels, Frankfurt and Hamburg, but its services were suspended after only six weeks and it missed out on any largesse. Suckling Airways was given approval for Manchester–Ipswich–Amsterdam with, interestingly, its seventeen-seat Dornier 228.

## Night Operations

> *This is the Night Mail crossing the border,*
> *Bringing the cheque and the postal order,*
> *Letters for the rich, letters for the poor,*
> *The shop at the corner, the girl next door,*
> *Pulling up Beattock, a steady climb,*
> *The gradients against her, but she's on time.*

The rhythmic chugging of the overnight mail train and the romance of the age of steam in W H Auden's wonderful poem happily is kept for posterity in Doctor John Grierson's classic GPO documentary. It became a distant memory from 1980, however, when aircraft took over much of the Royal Mail's night operation.

During the 1980s, the Post Office developed a countrywide, comprehensive network of night operations with many small aircraft operating seventy mail flights every night to major bases at East Midlands, Luton, Liverpool (Speke), as well as Glasgow, Edinburgh and Aberdeen. Around half of the Royal Mail flights served the GPO's major hub at East Midlands Airport with its purpose-built hangar and chief sorting office. The nightly operation had to run like clockwork and had to meet strict deadlines for its four million first class letters, and packages weighing up to 27.5 kilograms with Datapost guaranteeing same day/next day delivery anywhere in the UK. The activity was frenetic at the Post Office hubs. Speke Airport in Liverpool was the second busiest hub after East Midlands and handled some 6,000 bags every night, a number which could increase to 10,000 around Christmas. Each plane was loaded, unloaded and turned round in an hour. This was hard, heavy work. Plenty of activity and movement, but perhaps lacking the rhythm and certainly lacking the romance, to inspire a latter-day W H Auden.

For a long time Loganair had looked covetously at Air Ecosse's extensive night mail operations for the Post Office with its fleet of Bandeirante aircraft painted in Datapost's livery. Early tenders by the Company for a share of this work had been unsuccessful, but in 1980 the Post Office awarded the Company two contracts. These were to carry letter mail between Luton and Liverpool with the Twin Otter, and Datapost consignments between Manchester and Luton with the Shorts 330.

Air Ecosse, however, continued to have a firm grip on the Datapost contracts. It was only when it got into financial difficulties and was having to dispose of some of its aircraft in 1982 that Loganair got more involved when subcontracted by Air Ecosse to allow it to fulfil its contractual obligations. Loganair was given a further opportunity when additional aircraft were employed during the National Rail Strike. Separately, the Company was also awarded two Twin Otter contracts to carry newspapers between Luton, Glasgow and Leuchars.

Many small regional airlines like Loganair were desperate to share some of this night work to supplement their daytime flying activities, even although tenders were set on a marginal-costing basis – the price intended to cover all direct operating costs together with a small profit margin. The night contracts afforded no scope to charge for the aircraft or Company overheads. For several years, Air Ecosse's business had been overly dependent on these marginal night operations alongside a very fragile daytime flying programme and this imbalance would be its undoing.

By 1983 Loganair's night operations involved five aircraft, two Twin Otters and three Shorts 360s, of which three were on behalf of Air Ecosse. This had increased to seven aircraft by 1986, four for the Post Office, one for Datapost and two newspaper contracts, all of them now operating in Loganair's own right. The five Shorts 360 aircraft used were ideal for the carriage of the mail in their square upright cabins. The Loganair engineering staff reconfigured these aircraft every night – thirty-six seats removed and plywood flooring and side panels fitted, all in twenty minutes – to carry up to three tonnes of mail. Month after month, year after year, the Shorts 360s, and to a lesser extent the Twin Otters, flew from Glasgow and Edinburgh to Liverpool, Luton or East Midlands.

For an airline which had historically struggled to achieve adequate annual utilisation for its aircraft in the Scottish environment, these night operations were like manna from heaven. Loganair was now achieving more than 3,000 flying hours annually for each of its Shorts 360s, made possible also by the CAA's approval of a progressive maintenance programme for the Company's Shorts 360 fleet in the

*A typical scene at East Midlands Airport. In the late 1980s, the Company regularly had up to seven aircraft, including five Shorts 360s, employed on night mail operations. Seats were removed every night and cabin interior protection fitted. The plywood side panels used can be seen through the aircraft windows.*

Courtesy of Loganair

Glasgow hangar which obviated the usual need to have periods of downtime for the aircraft for maintenance purposes.

Night operations became a way of life for many Loganair pilots, engineers and operations staff. Night operations were also a crucial source of income for the Company from Datapost or letter mail and by the late 1980s represented more than fifteen percent of the Company's turnover. It was therefore a severe blow in June 1991 when the Post Office carried out a review of its night operation network. Two of Loganair's Datapost contracts were discontinued immediately and all other contracts were to be re-tendered. Although the Company successfully tendered for new contracts from Glasgow and Manchester to Coventry, and from Belfast to Luton, the night mail activity was drastically reduced. And just at a time when everything else was going wrong for the Company - again.

### Real Route Expansion
In 1981, when British Airways Highland Division was set up and the CAA decreed that the Scottish Highlands and Islands routes would be 'off limits' for the foreseeable future for Loganair and the other airlines, the old Loganair Board had to look further afield for growth and new scheduled services. The Company's new services to Belfast from Edinburgh and Glasgow were becoming successful. On the Glasgow–Belfast route, quickly followed by Edinburgh, Loganair's switch from Aldergrove Airport to Belfast Harbour Airport in 1983 had given it a real competitive edge over British Airways' service. Being near the city centre, the airport was much more convenient for the business passenger than Aldergrove, seventeen miles away. It was also

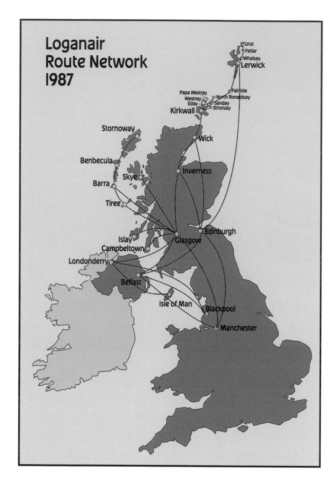

Loganair
Route Network
1987

*When the decision was taken in 1987 to bid for the former British Caledonian route licences for the Glasgow and Edinburgh–London Gatwick routes, Loganair's network was well-established and profitable. The double triangles linking Glasgow and Edinburgh with Belfast and Manchester were now the airline's principal routes.*

extensively used by the Ministry of Defence (MOD) for military personnel travel, especially when there was a Scottish regiment on tour in the Province. Belfast was now very important to Loganair.

Manchester was just about to become even more important. With the award of the Edinburgh–Manchester route licence in 1982, Loganair had gained access to its biggest market-place. Starting a new service in late 1983 from Manchester to Belfast seemed a natural development. When the Company had the audacity to win the licence to operate on the Glasgow–Manchester route the following year, British Airways was undoubtedly rattled.

Loganair's challenge now was to make a success of these services, especially the Manchester–Glasgow and Manchester–Belfast routes where British Airways was intent on aggressively retaining its market share. For its part, Loganair upgraded its services. The Shorts 330, which had inaugurated the Edinburgh–Manchester service, was replaced by a mix of Shorts 360 and Fokker Friendship aircraft, the service frequency was increased, and a better in-flight service offered. On the Glasgow and Edinburgh to Belfast routes, the Twin Otter and Shorts 330 were replaced by Shorts 360s.

The Company's management had every right to feel well satisfied at its string of successes at the many Licence Hearings before the CAA. At some of the early Hearings, the Company had presented its own case but, as the stakes grew higher, Loganair was represented by several eminent aviation lawyers including David Beety, Richard Venables and Robert Webb. At the most critical Hearing of all, relating to the ex-BCAL routes from Glasgow and Edinburgh to London Gatwick, Loganair was represented by Harvey Crush. Loganair's application was to be unsuccessful and Loganair would never be the same again.

### The Gatwick Route Licences

1986 had been a most difficult year for airlines and British Caledonian had lost £14.4 million. Worse still, prospects for the coming year were not encouraging for the airline. By then it had become clear that BCAL had not become 'the Second Force Airline' that the Edwards Committee report had hoped for back in 1969. Sir Adam Thomson realised his airline was being squeezed and a merger with another airline was his best option. Despite the constant strife between them for years, it was announced in July 1987 that there would be a merger, in effect a takeover, by British Airways. Before the deed was done, there was much haggling over the price. This was initially affected by the British Airways share price falling during the stock market crash, the Black Monday of 19 October 1987, and then further complicated by an approach being made by Scandinavian Airline Systems (SAS). There was a referral to the Monopolies and Mergers Commission (MMC) and, to ease things along, Lord King, British Airways' Chairman offered a number of concessions including 'giving up all BCAL's domestic and Channel Islands services from Gatwick'. These, of course, included the Glasgow and Edinburgh routes.

The Loganair Board had just finished its meeting in Glasgow and Michael Bishop took a telephone call from Ian Imrie, the redoubtable Industrial Editor of the *Glasgow Herald* newspaper: 'Would British Midland be applying to operate the ex-BCAL routes to Gatwick?' Michael Bishop explained that he would not be applying as BMA had, for several years now, been building up an impressive number of scheduled services at its Heathrow base, including, of course, the cross border services from Glasgow and Edinburgh in competition with the British Airways' Shuttle. Ian Imrie then asked, 'And what about Loganair?' There was a slight hesitation, 'Yes, they would be applying.' 'With which aircraft?' 'British Midland will cascade our DC9 aircraft to Loganair for the BCAL services.' The Loganair Board thus learned that Loganair was being catapulted from turbo-prop into the jet age.

This was serious policy-making on the hoof which would have disastrous consequences for Loganair. The Company, however, just had to get on with it, and the first vital step was Loganair getting into a position to demonstrate that it could in fact make the transition, indeed the quantum leap from turbo-prop to jet. Jet capability was essential if Loganair's application for these trunk routes was to be credible. Meanwhile, Michael Bishop had thought better of using the British Midland McDonnell Douglas DC9 aircraft. Instead, Loganair was to purchase two brand new BAe 146-200 one hundred and one-seat jets. There was a certain logic to this change. Manx Airlines already had a BAe 146-100 in its fleet for the short runway at Ronaldsway Airport, and Eurocity Express, a fellow subsidiary and soon to be renamed London City Airways, had plans to replace the de Havilland Dash 7 with the BAe 146, when permitted to do so, at the new Docklands STOLport. The BAe 146 was to become an important aircraft for the Group Companies. Purchase Agreements for Loganair's two aircraft were quickly signed with British Aerospace. The aircraft were on their way. Fingers crossed, of course, that Loganair's licence applications for the two ex-BCAL routes would be successful and Loganair would actually have routes on which to use them.

*The BAe 146. The Company's two 'Whispering Jets' were purchased specifically for the former British Caledonian routes from London Gatwick to Glasgow and Edinburgh. Unfortunately the route licences were awarded to Air UK. The 101-seat aircraft was deployed mostly on scheduled services from the Company's Manchester hub, but was also used for holiday charter work.*

Artist: Loganair pilot
R N Jamison

Harvey Crush was advising the Company at the Licence Hearings in June 1988. Other applicants for the Glasgow–London Gatwick route were Air UK and BIA; and for the Edinburgh licence, there were three other applicants, Air UK, BIA and Dan-Air. Loganair had strong Scottish support for the Gatwick applications and the Company's witnesses at the Hearing included Ewan Marwick, Chief Executive of Glasgow Chamber of Commerce, Colin Carnie, Area Chairman of Scottish Council Development & Industry, and Roy Brabender, President of the Scottish Passenger Agents Association.

The Company had earlier submitted its Business Plan and all supporting financial information to the CAA and there appeared to be no concerns about the Company's financial fitness. Loganair's case on paper appeared stronger than that of the other applicants with its better frequency, timetable and route economics. Importantly also, Loganair had the aircraft in place to do the job. Two BAe 146 jets had been acquired and one was already operating with Loganair. To the great credit of Loganair's management, pilots and engineers, the Company had made the transition to jet equipment in a very short time and was indeed a credible contender. It had every right to sit at the big boys' table.

It is fair to say that, from the outset, Harvey Crush had serious misgivings about the Company's chance of success at the Licence Hearing. His essential concern was the obvious connection between Loganair and British Midland, both now wholly-owned subsidiaries of the new Holding Company in the reconstructed Group, Airlines of Britain plc. Michael Bishop, of course, was Group Chairman, Chairman of British Midland and Chairman of Loganair. British Midland was already a major operator on the Scottish trunk routes to London Heathrow. It was well understood that the

*The BAe 146 and the ATP were a common sight at Manchester Airport. As well as scheduled service routes to Glasgow, Edinburgh, Belfast, Londonderry and Dublin, the Company also operated holiday charters and in 1989 was the second biggest operator at Manchester Airport.*

Photo: British Aerospace

Government was anxious to develop effective scheduled services to London Gatwick. The question of arm's length competition between British Midland and Loganair was always going to be the central issue and Harvey Crush's early apprehensions were increased when the other applicants' Statements in Objection unsurprisingly majored on the close relationship between Michael Bishop's companies.

Michael Bishop wished to appear personally at the Hearing as Loganair's witness and it was only with the greatest reluctance that he was persuaded to remain out of sight. Distancing him and British Midland from Loganair was vital. The author made the case for the Company during several hours of grilling by a Panel which always reverted to the same central issue. He stated repeatedly that it was Group policy to give autonomy to the individual companies within the Group, to Manx Airlines and Eurocity Express as well as to Loganair. It may well have been policy, but as the Chairman of the Panel, Mr Raymond Colegate, affirmed such policy could always be changed if Loganair became a successful Gatwick operator. Effective competition with British Midland at Heathrow, and real competition between London Gatwick and London Heathrow airports, would not be guaranteed if Loganair was awarded the Gatwick licences irrespective of any policy of autonomy within Airlines of Britain.

Industry observers were not surprised when the route licences for Glasgow and Edinburgh to London Gatwick were awarded to Air UK. Michael Bishop continued to believe his presence at the Hearing would have made a difference to the outcome. He should not have been talked out of it. He remained very resentful. The CAA decision was clear enough, 'The ownership and control of Loganair is such that the Authority cannot believe the airline would be allowed to undertake any activity which, while benefiting Loganair itself, was a substantial net disbenefit to the Group as a whole.' So that was that. Loganair had two hugely expensive BAe 146 aircraft for which their intended task had just disappeared.

*There was great excitement at Glasgow Airport when the Company's first ever jet arrived. The engineering and flight operations management were complimented on getting the aircraft on to the Company's Air Operator's Certificate by the date of the CAA Route Licence Hearing in June 1988.*

Courtesy of Loganair

## Manchester Opportunities

The Glasgow and Edinburgh to Gatwick routes, each with their annual passenger traffic of more than 150,000, would have elevated Loganair into a bigger league. Alas, it was not to be, and the Loganair management immediately set about establishing a new market for the BAe 146 aircraft, in particular the holiday charter market. In the winter of 1988/89, the Company did extremely well to secure a comprehensive ski holiday programme. Although the decision had been taken not to fly to the difficult Innsbruck Airport until the Company's pilots were more experienced on type, weekend services were operated from Manchester and Gatwick to Munich, from Glasgow and Birmingham to Geneva, and from Manchester to Salzburg and Nice. The following summer, the BAe 146 was heavily engaged in Mediterranean inclusive tour work from Manchester to Palma, Malaga, Alicante, Murcia and Ibiza.

Dancers welcome Loganair's BAe 146 at Murcia Airport. The two 146s were used on the scheduled services and on holiday charters from Manchester. There were winter ski charters to Munich, Geneva, Salzburg and Nice, while there were summer flights to Mediterranean sunspots such as Palma, Ibiza, Alicante and Murcia.

Courtesy of Loganair

Several series of charter flights were secured: to carry RAF personnel from Manchester to Gütersloh for the MOD for more than a year; for Nortjet operations in San Sebastian and Majorca for several months; and also for British Airways Highland Division and Aer Lingus. The best prospect was a joint venture with Sabena for scheduled services on the Manchester–Brussels route, just as British Midland had with Sabena from Birmingham to Brussels, and London City Airways operated from Docklands STOLport to Brussels. Disappointingly, the Company's joint venture with Sabena on the Manchester–Brussels route was short-lived. After SAS invested in the ABH Group, great efforts were being made to identify economies and synergies through collaboration. Loganair approached SAS about a joint arrangement on the Manchester–Dublin route, but because of difficulties with the SAS unions, nothing ensued. On the scheduled service front, the BAe 146 were used on the recently licensed leisure routes from Manchester to Guernsey and Jersey, but were primarily deployed on the services from Manchester to Edinburgh and Belfast.

From a standing start, the Company had done well to secure so much new business for the BAe 146 aircraft, much of it from Manchester. Manchester Airport's catchment area, with its population of twenty million, was huge by comparison with anywhere else Loganair operated and had to be key to the Company's much needed growth. Much time, effort and money was now being invested in building up the Company's presence at the airport, together with significant advertising and marketing investment in promoting the Company's charter and scheduled services. In May 1989, Loganair launched a new brand, *Premier Service*, and opened a *Premier Service* frequent flyer lounge in the new domestic terminal at Manchester Airport. The Chairman stated in the Airlines of Britain Group Accounts of 1989 that, in terms of number of flights, Loganair was the second biggest operator at Manchester Airport.

It was during this time that the Company reached another milestone – 250,000 passengers in the year. By careful calculation, it was found that the 250,000th passenger would be on the morning Manchester–Glasgow flight on 13 December 1987, and the Company's Public Relations manager, Roger Palmer, arranged to

Courtesy of Loganair

*Premier Club Lounge.*
*The new domestic terminal at Manchester was opened in June 1989 by HRH the Princess of Wales. At the same time, Loganair introduced its Premier Service, the new branding for Loganair's flights from the Manchester hub. The Premier Club lounge facility for frequent travellers was also opened.*

present the lucky passenger with a bottle of champagne with the mandatory newspaper photographer on hand. Just in the nick of time, Roger Palmer noticed that his chosen winner was actually handcuffed to the man standing close by him in the queue. With commendable nimble footedness, he moved forward and a bottle of bubbly was duly presented to Loganair's 249,999th passenger instead! Loganair services, especially to the islands, regularly carry prisoners along with one or even two escorts. Crime really does pay for the Company.

At this time, Manchester International Airport (MIA) was actively promoting a synchronised hubbing concept and had twenty-six user airlines involved, without British Airways which was perceived to be developing its own network hub at Birmingham. MIA was planning to financially encourage airlines to step up their service frequency at Manchester. To explore possible opportunities for Loganair, negotiations took place with Swissair, Sabena, Alitalia and Lufthansa. Loganair concluded an interline agreement with American Airlines to connect with its Boeing 767 Chicago–Manchester service, and the Company's flights appeared on American Airlines' Sabre CRS display with an 'AA' designator. The Company tried hard to capitalise on its strong position at Manchester Airport. It added a new scheduled service to Knock after Ryanair withdrew from the route and it finally started a service to Dublin, but without any tie-up with SAS. Later, a service to Londonderry was launched. The Company management was doing everything possible to make a financial success of its new aircraft operations, but it was to prove to be an impossible hand to play.

*Eye-in-the-Sky – Loganair's next aircraft for Barra? By the late 1980s Loganair was the second busiest operator at Manchester Airport with 150 flights per week. To help promote its many flights and Premier Service brand, the Company sponsored Piccadilly Radio's Eye-in-the-Sky traffic service for Greater Manchester. The four-seat Cessna 172*

Paul Francis Photography

*sported the Loganair livery. Piccadilly Radio was the biggest radio station outside London's Capital Radio and claimed to reach an audience of 2,950,000.*

## Fleet Changes

By the time the second BAe 146 aircraft arrived, the Company had been in modest profit for five consecutive years. The two jets had been acquired for two specific high density tasks – BCAL's Glasgow and Edinburgh to Gatwick had annual passenger traffic of 146,000 and 150,000 respectively. When the licence for these routes failed to materialise, the Company's management had to make the best of the situation by using the BAe 146 aircraft as sensibly as possible on its existing routes and finding additional work wherever possible. It was a huge task which was about to get even more daunting, for the Group meanwhile was intent on acquiring more new BAe ATP aircraft to be used by British Midland, Manx Airlines and Loganair. A policy decision had been taken at Donington Hall, British Midland's headquarters, to replace with ATPs, the last remaining older aircraft in the Group and these included the three Fokker F27 aircraft being used by Loganair. From the Group perspective, this was progress. From Loganair's standpoint, it meant that some relatively fragile scheduled services, operated by three elderly and cheap Fokker F27 aircraft, would now have to bear the significantly greater costs of the new BAe ATP.

The Chairman acknowledged, in 1989, that the Group's decision to modernise the aircraft fleet would impact severely on Loganair. In the following two years, however, with Loganair's mounting losses and severe cash flow difficulties, this essential underlying cause of Loganair's acute problem would be quickly forgotten. Having been stranded with two jet aircraft which were too big and seriously too expensive for Loganair's normal type of work was bad enough, but now having the additional burden of the BAe ATP aircraft foisted upon it made the task of the Company's

*Loganair's first sixty-four seat ATP was delivered in 1989. By 1993 Loganair was operating five of the Airlines of Britain Group's ten ATP aircraft when it ordered three more ATPs and the Group ordered another six ATPs for delivery in 1994 and 1995 – all as part of the BAe 146 Exit Agreement.*

Photo: British Aerospace

management well nigh impossible. Just when it appeared that things could not get worse, the Chancellor of the Exchequer raised bank interest rates. During negotiations for the BAe 146 aircraft and the Group ATPs, the London Interbank Offered Rate (LIBOR), the usual interest rate used in aircraft negotiations, had been 8.5 percent. This had quickly increased to fourteen percent.

To the author's dismay, the Company's two ATPs suddenly became three when John Wolfe announced that British Midland had no need of one and it was being transferred to Loganair. To make matters worse, the Group's ATPs were all late being delivered while the three Fokker F27 aircraft that Loganair had been using had meanwhile been sold by British Midland. Loganair therefore had to lease in an assortment of aircraft to maintain its scheduled services: a Ryanair BAC 1-11-500 followed by a Ryanair 1-11-400; a Manx ATP for weekday services; and a Fokker F27 from Busy Bee in Norway followed by a Fokker F27 from Star Air, a subsidiary of Maersk. These were difficult times for the Company's Operations Department and, despite everyone's best endeavours, the scheduled service programme was severely disrupted which simply made matters worse. Loganair's first ATP was eventually delivered in September 1989.

### British Airways Highland Division

Never far away from all that was going on in Loganair was British Airways Highland Division. 'Highland Division' was actually a misnomer. Certainly its *raison d'être* was to provide services in the Scottish Highlands and Islands, but it also had cross-border services and an operation from Berlin. Highland Division, in effect, was British Airways' turbo prop division. Nor was it ever far from the minds of the Loganair management that BAHD was keeping the main routes in Scotland frustratingly just beyond its reach. While Loganair had to make do with the thinner routes, BAHD services included all the high density routes in Scotland, and still Highland Division's operations were marginal at best. Clearly Loganair's network would have been bolstered by access to some of BAHD's routes without, as Loganair would always argue, making any difference whatsoever to British Airways in the greater scheme of things. In the early 1970s, there had been uncertainty about British Airways' intentions in Scotland. In the mid 1970s, there was uncertainty about just how many of its Scottish routes it would divest. In 1980, it had been a racing certainty that it would be withdrawing altogether from Scotland; and in 1984 the CAA Report had been expected to recommend a realignment of British Airways' Scottish services. Despite all of that, British Airways was still there and little had changed in Scotland.

Forming Highland Division was hardly a panacea for the financial shortcomings of British Airways' Scottish services. As early as 1982, hardly a year after Highland

Division was formed, Hugh Reid, British Airways' Scottish Manager, was again looking for further economies. He discontinued the Aberdeen–Inverness link and, more confusingly, also proposed that BAHD would go back on to the Stornoway–Benbecula route from which British Airways had withdrawn seven years earlier 'in order to save £187,000 per annum'. Loganair was incensed as it had just negotiated a subsidy from the Western Isles Islands Council (WIIC) for a service with the bigger Twin Otter aircraft on the Stornoway–Benbecula–Barra route. For its part, WIIC was surprised as BAHD had made no mention of this intention at the recent British Airways Consultative Committee meeting held at O'Hare Airport, Chicago. For many years it was British Airways' practice to hold these meetings in faraway places like New York, Boston or Paris. Clearly Highland Division could offer more than anything Loganair could offer its customers. In the end BAHD did not resume services between Stornoway and Benbecula.

Highland Division was still fighting off all competition while the London management was trying in vain to use the newly-won revolutionary working practices elsewhere in its airline operations. That clearly was taking time and meanwhile

*Signing ceremony, Inveraray Castle. Loganair was competing with British Airways on several routes and was anxious to improve its inflight catering and service. In 1985, Loganair negotiated with Pernod Ricard, owners of Aberlour, Glenlivet and Campbell's Whisky, and became the first airline in the UK to offer malt whisky on board. Pernod Ricard had the Duke of Argyll, head of the Clan Campbell, on its Board and it used Inveraray Castle for promotional purposes. It was arranged that the Loganair Agreement be signed at the Castle.*

Courtesy of Loganair

*There was little conversation at lunch as the Duke was more content to smoke his cigarettes throughout until, he announced to the author, 'I regularly fly Loganair to Tiree.' Scott Grier proclaimed himself to be very pleased: 'And why do you go a lot to Tiree, your Grace?' 'Because I own it', was the reply. The author resumed his delightful conversation with the Duchess.*

BAHD continued to struggle to attain profitability. It undoubtedly had made great strides since it being formed in 1981 when it was losing £6.1 million per year. BAHD had achieved considerable economies, primarily through reduced crew costs and allowances, and also productivity increases of more than thirty percent. It was now a leaner organisation, but still not on an even keel.

Nor had British Airways won hearts and minds. A little more than a year after BAHD was formed, there was real impatience being expressed. At the CAA Tariff Hearing in February 1983, representatives of the Western Isles, Orkney and Shetland Islands Councils objected to British Airways' application for a fare increase significantly above inflation and they complained that BAHD was not passing on to the passengers the benefits of its much publicised cost-cutting. Indeed, Highland Regional Council went further. If British Airways could not deliver benefits, it should stand aside. Dan-Air was taking over from British Airways on the Inverness–London Heathrow route and was proposing more attractive fares and schedules. Air Ecosse had taken over from British Airways on the Aberdeen–Wick route and was providing three daily services compared with British Airways' former single service. Highland Regional Council 'viewed with equanimity the prospect of losing British Airways as other airlines were seriously interested in taking over.'

In 1984, BAHD's claim in the *Glasgow Herald* that it had made £1 million profit was met by general scepticism. If separate Highland Division Accounts were produced, they were certainly never in the public domain. Such a significant turn round in such a short time was seen by many to be unlikely, and indeed extremely unlikely if the more expensive HS748 aircraft, and later the new Super HS 748 which were being introduced to the Highland Division fleet, were being properly accounted for. Unfortunately for BAHD also, the real cost savings from the revolutionary work practices were relatively short-lived. From the time BAHD moved to Berlin to operate four HS748s on the Internal German services, and pilots' union BALPA had demanded that its members receive British Airways mainline allowances, all the cost savings and efficiencies had been eroded. Years later, in 1996, Bob Ayling, British Airways' Chief Executive, admitted: 'We have been flying on these routes in one form or another for the past fifty years, but throughout the period we have never achieved profitability.'

Nor did British Airways fare any better elsewhere from the BAHD experience. The British Airways management had high hopes that the Highland Division model could be a blueprint to be adopted in other parts of the operation, but history would show that this was simply not to be. The British Airways Trade Union Council could be proud that, by accepting these practices, many jobs in Scotland were saved, but they were heavily criticised by the British Airways union members at London Heathrow, especially by cabin crew who saw the Highland Division work practices as the thin end of the wedge.

It would be many years before the hoped-for gains in efficiency would begin to be achieved and these had nothing whatsoever to do with Highland Division. It was the 1990s before engineering staff complements were reduced. It took the *Guiding Principles* programme of 1995, then 9/11, followed by the dire Pension Fund deficit problem, before pilots accepted change and productivity was improved. It was only with the move to Terminal 5 at Heathrow that some of the old Spanish practices in ground operations were ended, and it was 2011 before any real progress was made in changing the cabin staff conditions of employment and only then after two years of industrial strife.

### The sleeping giant awakens

Despite British Airways' overwhelming strength in the marketplace, Loganair had made steady progress throughout the 1980s in developing its scheduled service network. Loganair, without question, was a serious player at Manchester and Belfast. It could not last. In 1989, British Airways was now flexing its considerable muscle and fighting its corner most aggressively, and competition with Loganair was bitter on the Belfast and Manchester routes for the rest of the decade. Loganair was hurting, particularly on the Glasgow–Manchester route.

Worse still, to Loganair's great dismay in 1989, British Airways was back on its precious Manchester–Edinburgh route. Having had the service to itself since it quite literally had seen British Airways off the route back in 1982, the Company now found itself in direct competition again. Loganair had not only won the licence in 1982, but had successfully revoked the British Airways route licence, or so it thought. Unbeknown to David Beety, the aviation lawyer who had presented Loganair's case, British Airways held a second licence covering the route. Ironically, it was only seven years later, when Loganair was going through the process of revoking the BCAL licences, that the surviving British Airways licence came to light. British Airways immediately decided to use it and, at the subsequent licence Appeal Hearing, neither the eloquent objections of Michael Bishop this time, nor of the author, were successful and British Airways resumed service, initially with a single daily frequency.

Fresh from this success, British Airways immediately applied for, and was awarded, a licence on the Edinburgh–Belfast route which it had unashamedly abandoned in 1980. Lord King announced: 'Against moves to block us by the Airlines of Britain Group, we have just broken the monopoly on the Belfast–Edinburgh route to give the customers the choice they did not have before.' There was a bigger agenda unfolding. There is no doubt that the British Airways backlash was against Michael

Bishop and his Group of Companies and the wider threat they posed. It would be Loganair, however, which would suffer. Suddenly, having new competition from British Airways on two of its most important routes was a serious double blow to Loganair. There was a further sinister threat as Lord King went on to proclaim: 'And we have news for them. We will be looking to do the same for more communities in this country.' Meanwhile, on the Glasgow–Manchester route, despite increasing frequency to three services per day, the Company simply could not increase its market share in head to head competition with the mighty British Airways. Loganair's losses were unsustainable and the service was terminated.

There was one interesting episode in this period of cut-throat competition on the Glasgow–Manchester route. In 1989, because its BAe ATP aircraft delivery had been delayed, British Airways had leased in a BAe 146 aircraft from Presidential Airlines and used the jet on the route to further undermine Loganair's efforts. This episode would go down in Scottish aviation folklore. Between Manchester flights, the BAe 146 sat motionless on the tarmac at Glasgow Airport until the British Airways management decided it also could be used on the Glasgow–Benbecula route. The Uists had never had, and would never have again, such a grand, quick service, but it was not to be without its critics: 'In the good old days of the Viscount, you had time for two free drams, now there's only time for one free dram.'

### Tit for tat in the Highlands and Islands
Now under severe pressure from British Airways on its services furth of Scotland, Loganair felt it had to retaliate and so set out to compete with BAHD within Scotland for the first time since it was formed. This time, the Monopolies and Mergers Report on the British Airways/BCAL merger the previous year made it difficult for BAHD to lodge an objection. Loganair was awarded licences in September 1989 to operate on the Glasgow–Benbecula, Glasgow–Stornoway and the Stornoway–Inverness routes using its 36-seat Shorts 360s against British Airways' 44-seat HS748.

From British Airways' perspective, re-entry on to the Edinburgh to Belfast and Manchester routes made commercial sense. No doubt British Airways also derived some satisfaction in having driven Loganair off the Glasgow–Manchester route. Its tit for tat reaction to Loganair's temerity in applying for the Western Isles routes, however, was less understandable. British Airways' licence application for the marginal Benbecula–Stornoway route, on which the Loganair service was dependent upon Council subsidy, was bizarre. The Council and Loganair did object and British Airways' applications came to nothing.

British Airways' next licence application, for the Inverness–Wick–Kirkwall route operated by Loganair's Islander aircraft, made no more sense as the Inverness–Wick service was even more marginal. When BAHD applied for a licence to compete, Loganair announced its withdrawal from the route. A somewhat embarrassed BAHD then withdrew its application. Early hopes that the route would be taken up by Gill Airways, or possibly Aberdeen Airways, the former Air Ecosse, came to nothing.

### *Away from British Airways*

The Company's preoccupation with British Airways and what it was doing, or what it might do in Scotland, fortunately did not close management's eyes to other commercial opportunities that presented themselves. When, on 27 June 1990, Yorkshire-based Capital Airlines suddenly stopped operating, the Company immediately applied for a licence to operate the Glasgow–Leeds and Leeds–Belfast services. The CAA acted quickly and awarded Loganair a three-month interim licence to operate on the Glasgow route with its 26,000 annual passengers. The Company was content to leave the Leeds–Belfast route to Jersey European. Having been granted the licence on Wednesday, the Company commenced a double-daily service from Leeds to Glasgow the following Monday, displaying a commendable nimble footedness. There had been hopes in Yorkshire that Capital, which had quickly expanded from being a fairly successful Shorts 360 operator to a BAe 146 jet operation, and had got into severe financial difficulties in the process, would somehow be resurrected. It was not to be and Loganair was subsequently granted a full licence. The Leeds–Glasgow service was operated successfully by the Company until 1993 when British Midland decided it would fit well with its Leeds-based operations. The service has continued to the present day under the British Midland banner.

Photo: David Dyer Collection

*In 1989, the Company purchased a hot air balloon from Cameron Balloons in Bristol to be used for marketing publicity and promotional purposes and also as a sporting interest for the staff. Promotional opportunities using the balloon were severely limited by windy conditions and the balloon was subsequently sold to an interested employee.*

## Difficult Times for the Industry and More Talks

In January 1991, the industry as a whole was in dire straits as a result of the continuing Gulf War accompanied by severe downturn in air travel and soaring fuel costs. British Airways acknowledged publicly that if the conflict dragged on, the airline would be prepared to shed 5,000 of its 50,000 staff. Meanwhile, Chief Executive Colin Marshall told staff they could go on extended, unpaid leave. British Airways certainly was not having its sorrows to seek. The granting of North American slots at Heathrow to Virgin Atlantic Airways, and the takeover by United Airlines and American Airlines of the transatlantic routes of Pan Am and TWA, certainly increased pressure on British Airways which was understood to make sixty percent of its profit from its transatlantic services. Elsewhere, job losses had already begun in airlines including British Midland, Air Europe and SAS. The situation certainly was serious. Most international airlines had been forced to cancel up to half their flights and wherever possible use smaller aircraft. British Airways had gone one step further by scrapping all flights to its three destinations in Ireland - Dublin, Cork and Shannon.

Nearer home there had been a serious fall-off of passenger traffic on all Scottish air services. British Airways was once again in the mood to discuss collaboration with Loganair. Joint operations would clearly permit substantial fleet savings, synergies and economies of scale. John Wolfe and the author met Rod Hoare, General Manager of British Airways' newly created Regional Business Units, and discussed three possibilities: Loganair take over Highland Division; British Airways acquire a minority holding in Loganair; or a separate joint venture company be created. Rod Hoare's preference was for a joint venture company and was prepared to include the Northern Ireland routes along with the Scottish Highlands and Islands routes with Loganair taking over four of the British Airways HS748 aircraft. Between the two Companies there could be a combined saving of eight aircraft. Michael Bishop, however, was adamant Loganair would only proceed to the next stage if British Airways' Manchester ATP routes were also to be included. There the discussions stalled. Plus ça change.

## A major initiative

British Airways' position and future intentions were causing real concern in Scotland. There were seemingly ambiguous messages coming from Lord King. He had ordered a comprehensive review of all British Airways operations with a view to improving profitability. 'Nothing is ruled out. No route is sacred,' he had stated. There was uncertainty, yet again, about British Airways' future plans for Scotland. Following the failure of the talks with British Airways, and with all of British Airways' known difficulties continuing, Loganair thought the time was just right to take the initiative and give Lord King some food for thought.

With a view to applying more pressure on British Airways and, hopefully, hastening its departure from the Highlands and Islands, Loganair entered an agreement with British Aerospace to lease five eighteen-seat Jetstream J31 aircraft. Three of these would be replaced later by three new twenty-nine-seat Jetstream J41 aircraft which was part of the sixteen aircraft deal that was being negotiated between British Aerospace and the Airlines of Britain Group.

With these Jetstream J31 aircraft Loganair announced it would mount competitive services against British Airways from Glasgow, Edinburgh and Aberdeen to Orkney and Shetland. The Company claimed that these services would 'revolutionise' air travel in the Highlands and Islands. Certainly the new non-stop services from the Central Belt to the Northern Isles would take time and tedium out of the passengers' journey which, with British Airways, still meant a stop in Aberdeen or Inverness on the way. Loganair's proposed services with the Jetstream 31 would mean a one hour and twenty-five minute journey, compared with about three hours with British Airways, giving Loganair a serious competitive edge at last.

The Glasgow–Kirkwall route licence was granted and a twice-daily J31 service started in September 1991, but the Company's application for services from the mainland to Shetland, yet again, proved problematical. Loganair, having had discussions with BP and some encouragement from Shetland, had opted for Scatsta Airport rather than Sumburgh for these services. The Company had taken a full page advertisement in the *Shetland Times* in September 1991 to explain that, with British

*Jetstream 31 over Culzean Castle. In 1991, Loganair leased five, eighteen-seat Jetstream 31 aircraft in order to operate non-stop services in competition with British Airways on routes from Glasgow, Edinburgh and Aberdeen to Orkney and Shetland. The Jetstream 31 proved highly successful on the Glasgow– Southampton and Inverness– Manchester routes.*

Courtesy of
British Aerospace

Airways using Sumburgh Airport, and Loganair using Scatsta, Shetlanders would not only have a choice of airline, but have two airports with different geography and different runway alignment, so that much of Shetland's weather-enforced air service disruption would be avoided. This would prove controversial as Loganair inevitably were caught up in the Shetland airport debate. Decisions on the Shetland licence applications were deferred several times, but the use of Scatsta for scheduled services was eventually refused by Shetland Islands Council. The Company had its 'wings clipped' by councillors who wanted Sumburgh to remain as Shetland's main airport and to protect jobs in the south mainland.

There was no disguising just how serious the failure to start the Jetstream 31 services to Scatsta from Aberdeen, Glasgow and Edinburgh was for the Company and it was viewed badly by Messrs Bishop and Wolfe. The Loganair management acted quickly to employ these Jetstream 31 aircraft on new scheduled services further afield. The Inverness–Manchester service was up and running, replacing a former Dan-Air service. A double daily J31 service was begun on the Glasgow–Southampton route, which had not been served since British Airways withdrew in the late 1970s, and was quickly followed by an Edinburgh-Southampton service. Loganair's 'Heathrow bypass' was a success and was upgraded later by Loganair with its Jetstream 41, and after that taken over by Manx's Embraer 145 jet aircraft. Loganair also simultaneously started a new Jetstream 41 international service from Southampton to Brussels and this too would eventually be operated by jet aircraft.

### And British Airways Retaliation – wise or otherwise

British Airways may have been suffering from the economic recession, the Gulf War, new transatlantic competition at Heathrow, and losing money on its Scottish routes, but in the words of Rod Hoare, it 'could not let Loganair think that we are just going to sit there like lame ducks' in the face of Loganair's Jetstream 31 initiative. Unfortunately for Loganair, the British Airways response was to set up a separate Scottish short-haul operation. An agreement had been reached between British Airways and its three main unions to reduce staff costs across the UK. Scottish Regional would be a 'dedicated profit-accountable unit'. It immediately announced an order for five BAe ATP aircraft for the Scottish routes to replace the HS748s – the forty-four seat 'Budgies'. British Airways was 'here to stay' was the message. To emphasise the point, Rod Hoare, and Ian Reid, British Airways' Scottish General Manager, stated their intention to increase frequency on existing services, open up new destinations, and most exciting of all, offer cut price off-peak fares – some now £22 cheaper – and all with the expensive sixty-six seat ATPs. It was certainly a bold, but highly improbable, route back to profitability on its Scottish network, which had been loss-making, to say the least, since the Second World War. It was not quite the response that Loganair was hoping for.

Loganair had been providing services on the Glasgow–Stornoway, Glasgow–Benbecula and Inverness–Stornoway routes since 1989. The Company had tried to provide complementary services to those of British Airways and was undoubtedly offering the passenger greater choice and convenience with a good day-return facility for the Glasgow-originating passenger on the Stornoway route and likewise for the Stornoway passenger to Inverness. Despite having been welcomed initially by WIIC, Loganair's continual sniping at British Airways was causing tensions which were most apparent in the Western Isles. The Company's unpressurised Shorts 360 aircraft were unpopular and compared unfavourably with British Airways' HS748. What upset the Council even more, however, was that Loganair's thirty percent market share on the routes was hurting British Airways at a time when there were concerns about British Airways' future commitment to the routes. For many people in Scotland, and perhaps especially in the Western Isles, British Airways was still regarded as the 'state carrier' with a duty to serve the Highlands and Islands. The Loganair services were seen as spoiling tactics and there was bad feeling towards the Company.

With the continuing effects of the recession, achieving a profit was very difficult for both British Airways and Loganair, and the actions of both airlines smacked of desperation. It was agreed that the author discuss a trade-off with British Airways whereby Loganair would come off the Benbecula and Stornoway routes if British Airways reduced frequency on the Manchester–Edinburgh route. There would be no deal. What British Airways did next on the Manchester–Edinburgh route, however, was a real shock. British Airways increased its daily rotations on the Manchester–Edinburgh route from one to four, as well as introducing hugely discounted fares. This was intended to be, and most certainly was, hugely damaging to Loganair.

British Airways' action, however, proved to be ill-considered. Loganair believed that British Airways was employing anti-competitive practices on the Manchester–Edinburgh route. Hugh O'Donovan of Wilde Sapte was instructed by the Company and an application was made to the CAA under Section 24 of Licensing Rule 19 regarding predatory action. Certainly British Airways' much-increased frequency was excessive, was to the severe detriment of Loganair, and was likely to destabilise the Company and its Highlands and Islands operations. The CAA ruled after a preliminary Hearing – the first under this 'fast-track' procedure – that British Airways' frequency should be halved from four to two flights per day. The allegation was also made that British Airways had acted on commercial information available to them only under a Confidentiality Agreement. Before the date of the full Hearing,

Courtesy of Loganair

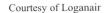

*Before it became politically incorrect, Loganair held beauty competitions for female staff members. Miss Loganair 1985, Karen Castle-Mason.*
*Miss Loganair 1989, Maureen O'Connor, and runner-up, Anna Hotten.*
*Miss World Airlines 1989. The winner Maureen O'Connor, Miss Loganair 1989, went on to compete in Paris for the Miss World Airlines title. She came second to Miss Air India while Miss South African Airways was placed third.*

British Airways climbed down and requested 'an indefinite postponement' and duly reduced its service frequency on the route back to one rotation per day.

British Airways was having a bad run. Now deemed guilty of anti-competitive behaviour against Loganair, very shortly it humiliatingly would have to admit being involved in a 'dirty tricks' campaign against Virgin Atlantic and be fined for its efforts. To industry observers, it seemed that British Airways had not learned lessons in the eight years or so since allegations of sharp practice had caused it to reach an $8 million settlement with Sir Freddie Laker.

Despite British Airways' retaliation with its announcement of its major investment in new BAe ATP aircraft, and the bluster that went with it, British Airways' intentions for its Highlands and Islands routes still remained ambivalent. Some observers, including Comhairle nan Eilean Siar Transport Chairman, Alex MacDonald, were convinced that British Airways' significant investment in a new fleet of upgraded BAe ATP aircraft was 'proof of the Company's continued involvement in the Islands.' Others were not so sure. Within six months, Brian Wilson MP, Labour's Scottish transport spokesman, had met British Airways' representatives in London and sought assurances following reports that the airline was considering hiving off its Scottish routes to Birmingham European Airways, an airline in which it already had a forty percent stake through its current Joint Venture with The Plimsoll Line. Eventually, the British Airways official line was: 'It is true that the routes are again in the red and we are looking at ways through joint consultations with our unions to bring them back into profit.' Hardly reassuring.

### Short-sighted decision or what?

Quite apart from its new services from the mainland to the Western Isles, and the constant difficulties with British Airways, Loganair's own critical financial position was putting the management under great pressure. Loganair had been providing services for the Council since 1975 on the Long Island route. A Britten-Norman Islander was used for the first three years on the Stornoway–Benbecula–Barra route, then was followed by the sixteen-seat Trislander before being replaced by the Twin Otter. Now, because of the Company's serious financial difficulties, and because it was under pressure to generate much needed cash from the sale of Twin Otters in the fleet, the author had proposed a reduced Long Island service to the Council. He was trying to make a virtue out of a necessity as passenger traffic was down due to the economic recession, and more especially because of economy measures to cut Council staff travel, taken after the Council's loss of £23 million in the Bank of Credit and Commerce International collapse. The proposal, effectively to operate the Glasgow–Barra service and the Stornoway–Benbecula–Barra service with a single aircraft was undoubtedly a compromise with the only merit being that it would have saved the Council from paying subsidy of £100,000 and would have allowed Loganair to sell one Twin Otter.

Loganair's innovative proposal was very badly received, and with emotions already raised by Loganair competing with British Airways on the mainland routes at a sensitive time, the *Stornoway Gazette* reported that 'the Loganair boss was pelted with verbal tomatoes by the Transportation Committee.' It was certainly true that the author did not need to wait for the English translation of the Gaelic used in the Council Chamber to realise just how angry some of the Councillors felt.

The timetable change on the Long Island route was a compromise proposed by the Company *in extremis*. The air service was a vital link as the overland route was such a poor alternative. The *Stornoway Gazette*, in March 1992, described it as:

> …tackling an obstacle course of ferries – not all of them vehicle carrying, some with no linking buses, many operating on alternative days, all changing from winter to summer timetable – as well as causeways, and a public transport network which is often sporadic.

It was little wonder that the WIIC Transportation Committee was so scathing about Loganair's plan. Needless to say, Loganair found another, more acceptable way to operate the Stornoway–Benbecula–Barra route for the Council.

### Meanwhile with British Aerospace

Throughout the time that Loganair was slugging it out with British Airways, the Company was wrestling with the operational, maintenance, and not least of all, the financial implications of having taken on new aircraft types, all in a very short space of time. It seemed a long time since the BAe ATP order for the Group Companies had been signed at the Paris Air Show in 1985. The order undoubtedly suited both parties admirably. For British Aerospace, it was most helpful having a prestigious UK launch customer as sales orders for the aircraft, to say the least, were sluggish; for British Midland having an effective trade-in for its fleet of aging Viscount aircraft was most opportune. There was a long lead time with the ATP order, but not so for Loganair's BAe 146 jets. Once Michael Bishop had decided that Loganair would make an application for BCAL's Gatwick routes, and had opted for the BAe 146 instead of his earlier choice of McDonnell Douglas DC9s, it was essential that Loganair took delivery of the aircraft quickly – and certainly before the Licence Hearings – to demonstrate to the CAA that the Company was a fit, qualified and a credible successor to BCAL.

Time was short and perhaps not surprisingly, with all the changes of mind, it seemed that British Aerospace was unclear about the Group's intention to acquire one, or perhaps two, or even three aircraft, all depending upon the outcome of the Gatwick

Hearings. Inevitably, there was some 'misunderstanding' about where exactly the Loganair aircraft were on the production line. It would take Sydney Gillibrand, the Deputy Chairman of British Aerospace, in person, having been summoned to Donington Hall, to sort out the problem and to ensure that one BAe 146-200 would be delivered to Loganair in time to meet its urgent requirement.

It is fair to say that, despite British Aerospace's best endeavours, the relationship with the Group was never easy over the following few years. Firstly, the ATP deliveries to the Group Companies were delayed, and in the interim, Loganair had to lease in an assortment of aircraft to maintain its scheduled service programme. Claims were made for compensation for costs incurred by the Company due to these delays. The Company was duly awarded substantial compensation for its two delayed ATPs, but the Loganair Board never accepted that the compensation was adequate. To make matters worse, Loganair's second BAe 146 was also delivered late, having suffered a serious bird strike prior to delivery. Further compensation was negotiated for this delay.

In the context of Loganair's still relatively small operations, the 101-seat BAe 146 and the 64-seat BAe ATP were expensive aircraft. The switch from second-hand, hand-me-down McDonnell Douglas DC9 to new four-engine British Aerospace BAe 146-200 may have had merit from a Group perspective, but for Loganair it meant a huge investment. The two aircraft cost £12 million each. Including two spare engines and rotables amounting to £3.2 million, and tooling and consumable spares, the total investment was well in excess of £30 million. Loganair's total annual turnover in 1988 was a mere £22 million.

A series of interest rate hikes by the Chancellor of the Exchequer meant that by the time Loganair's 146 and two ATP aircraft had been delivered in 1989, LIBOR was fourteen percent up from the 8.5 percent which had been assumed when the aircraft deals had been signed with British Aerospace. The Group by now had eight ATPs prompting John Wolfe to start negotiation for a restructuring of the leases for all Group aircraft. The new ATP monthly rentals were particularly serious for Loganair as the aircraft, including a spare engine and rotables, now costing the Company some £97,000 per month, was replacing the Fokker Friendship which had been costing £20,000 per month. Nor was the bigger aircraft expected to increase passengers or revenue in any significant way. Falling heir to British Midland's redundant ATP, Loganair's third, now added a million pounds a year to Loganair's cost base – unsurprisingly this was to the chagrin of the Loganair management.

The high cost of the ATP was bad enough. To make matters worse, the aircraft was experiencing serious technical difficulties which gave rise to an unacceptable incidence of unserviceability and service disruption. 'ATP' soon stood for 'Another

Technical Problem'. British Midland and Manx, and British Airways especially on its Northern Isles services, were also having similar difficulties.

Much worse still, there was also a serious loss of passenger confidence in the ATP over concerns about the aircraft's safety. Orkney and Shetland MP, Jim Wallace, sought assurances after a string of high profile incidents – they were guaranteed to be high profile after one incident involved an ATP carrying sixty-six mainly journalist passengers who were covering the Braer oil spill. Perhaps the most serious incident occurred when a British Airways ATP was caught in a crosswind on take-off from Sumburgh Airport, clipped its wing on the runway and was forced to make an emergency landing at RAF Kinloss. Peter Black, Head of Corporate Marketing and Public Affairs of Jetstream Aircraft Ltd, was sent to Orkney and Shetland to restore confidence. A modification programme affecting the Group's eight ATPs was devised, but unfortunately was much delayed and the Group negotiated a rebate on the monthly rentals.

### Farewell to the BAe 146 - but at what cost?

Meanwhile the BAe 146 was also causing serious concern as its four Textron Lycoming ALF502 engines were proving unreliable and costly. Other operators of the 146 aircraft, including Manx, Dan-Air, Air UK and the Queen's Flight, also had a high incidence of engine removals and Loganair started negotiating for compensation from Textron, this time in vain. Just as he had done with ATP monthly rentals, John Wolfe started similar negotiations to restructure the rentals of the two 146 aircraft, with the lease period extended from five to ten years.

Throughout this period, despite the enormous efforts of the Loganair management, utilisation of the 146 aircraft was never near adequate. By late 1991, the shortfall on

*The Roll-out of the Jetstream 41 at British Aerospace's Prestwick factory on 27 March 1991 in the presence of Her Majesty the Queen and HRH the Duke of Edinburgh was a spectacular extravaganza – very much the brainchild of Allan MacDonald, Director & General Manager of Jetstream Aircraft Limited.*

Courtesy of British Aerospace

the 146 operations was around £3 million. Enough was enough, and serious negotiations were started with Charles Masefield of British Aerospace to devise a mechanism for Loganair to extricate itself from the 146 contracts. Eventually an agreement was reached. Loganair's two 146 aircraft were to be returned to British Aerospace. The quid pro quo was that the Group Companies would commit to taking no less than sixteen additional aircraft: Loganair was to take five ATPs and three J41s, and Manx was to take eight aircraft. British Aerospace was to write off one-sixteenth of Loganair's deferred rental shortfall with the delivery of each new aircraft to the Group. Loganair was thus relieved of the immediate worry of the hugely onerous 146s, but with this sixteen-aircraft order, there would be a significant price to pay downstream.

Meanwhile, the Company found that even returning the two BAe 146 jets came at considerable cost through substantial legal fees, onerous aircraft return conditions, escrow and other capitalised aircraft introduction costs all having to be met or written off, all of which was fortunately eased to a degree by British Aerospace agreeing to buy Loganair's spare 146 engine. An expensive end to what had often been an exciting, but always extremely worrying, period in the Company's history, and all emanating from that abortive attempt to take over BCAL's routes from Glasgow and Edinburgh to Gatwick.

There was time for one last out-of-the-ordinary task for one of the 146 aircraft before its return to British Aerospace. Sir Michael Bishop arranged for it to be used on the successful campaign tour of Conservative Prime Minister, John Major, in the 1992 General Election. Later, the Loganair 146 crew accompanied Sir Michael to No 10 Downing Street to receive Mr Major's grateful thanks.

*For his election campaigning around the United Kingdom in May 1992, Prime Minister John Major used Loganair's BAe 146. After his successful election, Mr Major wrote to all staff to thank them individually for 'their quite outstanding service...after a long tiring day it was very comforting to see such friendly faces.' He also invited a number of the staff to accompany Sir Michael Bishop to a cocktail party at 10 Downing Street.*

Courtesy of Loganair

## The aircraft saga continues

For Loganair, this was the end of the 146 struggle, but not the end of the story. It was merely the start of seemingly endless meetings and negotiations with British Aerospace. The Group immediately challenged the cost of the monthly rentals for both ATP and J41 and reductions or deferred rentals were agreed. Integration funding from British Aerospace for each of the Group's eight Jetstream 41 aircraft was also negotiated.

ATP reliability, however, was continuing to be a concern to operators Manx, British Airways and British Midland, as well as Loganair, and this, together with further delays in Jetstream 41 deliveries, caused the Loganair Board to believe that further compensation was appropriate. Manx carried out an investigation into the number of ATP incidents and the aircraft's overall performance. As a consequence, the Group considered, at worst, returning all the Group ATP's, or perhaps only the three original prototype British Midland aircraft. Overarching these operational concerns was the view that British Aerospace production was in disarray and its future ATP and J41 production by no means certain. In view of the ten ATP aircraft now in the Group's operations, the Chairman sought clarity and reassurance about future production and product support from Sydney Gillibrand, Vice Chairman of British Aerospace.

Due to restructuring within British Aerospace, regional aircraft manufacturing was now the responsibility of Jetstream Aircraft Limited (JAL) at Prestwick, but it

Courtesy of British Aerospace

*At a joint launch ceremony, Sir Michael Bishop accepted the first Loganair Jetstream 41 and the first Manx Jetstream 41 from Michael J Turner, Chairman and Managing Director of British Aerospace Regional Aircraft Limited on 25 November 1992. Loganair would later claim to be the first airline to have the new aircraft in passenger service – on the Manchester–Southampton route.*

mattered not. Claims from the ABH Group and negotiations continued just the same. In summer 1993, Group negotiations were intense with Allan MacDonald, JAL's Managing Director, who endeavoured against the odds to meet the needs of his customer airlines. Loganair's cash flow problems were critical. JAL agreed to purchase from the Company and lease back three of the Shorts 360 aircraft with reduced monthly rentals, and also to take over the lease of Loganair's unsold spare 146 engine. Also, in return for a commitment to acquire three more ATP or Jetstream 41 aircraft, JAL agreed to delay various aircraft deliveries. However, matters were further complicated by seven J41 aircraft of Manx and Loganair requiring to be returned to Prestwick for modification, which meant transitional cover had to be arranged and this, of course, had to be paid for.

Throughout these years, British Aerospace had tried to be as accommodating as possible to the ABH Group which was, of course, one of its major customers. The aircraft manufacturer's position was weakened by serious aircraft delivery and performance difficulties. For several years it had been one problem after another, followed by one claim after another. Most of all, the lack of sales of both ATP and J41 created genuine concerns about future production. This would turn out to be prophetic. The BAe ATP manufacturing at Woodford and Chadderton was transferred to Prestwick for a virtual relaunch of the ATP as the Jetstream 61, but this had no more

Courtesy of British Aerospace

*Loganair took delivery of the first of its three Jetstream 41 aircraft in November 1992. It went into service on several Manchester routes and also the new international service from Southampton to Brussels. The J41 was the development of the eighteen-seat J31 built at the British Aerospace factory at Prestwick, but was less successful than the earlier aircraft. In 1997 production was stopped at the Prestwick factory with only one hundred J41 aircraft in service.*

success. ATP manufacturing ceased in 1996 with only sixty-four aircraft built. The following year, 1997, with 100 Jetstream 41 aircraft in service, British Aerospace announced it was stopping production at Prestwick. These two aircraft types – the last civil aircraft built by British Aerospace – had been a costly experience. As always, the lawyers had done well. The airlines – certainly Loganair – had not.

### Cash Flow

When the Company was acquired by British Midland in December 1983, Loganair had been operating with a bank overdraft facility and a term loan of £2.3 million from the Royal Bank of Scotland. In the later years of the 1980s with the new aircraft programme, each time a new aircraft was acquired, significant deposits, £200,000 for each BAe 146, and £75,000 for each ATP, had to be placed with the manufacturer. This further strained the Loganair overdraft facility already under pressure due to increased aircraft operating costs and income shortfall. Not surprisingly, the Company was making formal application on a regular basis for an increased bank overdraft facility. By the early 1990s, Loganair's account had been taken over by the Special Lending Department of the Royal Bank – the Company was now deemed to be in 'intensive care'.

Despite the many successful claims on British Aerospace which eased Loganair's cash flow, and despite sales of Twin Otter aircraft, the Company's overdraft requirement increased. By 1992, the Company's forecast cash flow requirement significantly exceeded the £2.5 million overdraft facility and had necessitated borrowing £2 million from British Midland and £750,000 from Manx to ensure that the Company did not breach its covenants entered into when the BAe 146 aircraft were acquired. For many months, British Midland routinely transferred funds for several days at the end of each month in a 'bed and breakfast' arrangement to allow the Company to stay within limits. The pressure from the Bank was eased, at least for the time being, by the sale and lease back of the three Shorts 360 aircraft. It was clear, however, that the underlying cash flow was deteriorating and the Board now considered seriously the Company's future.

### Management Buy Out

These were extremely worrying times. The continual cash flow difficulties and negotiations with the Bank were naturally a real concern to the Board. Moreover, the prospects for the Company were anything but auspicious. In December 1992, Rod Hoare had confirmed that British Airways would not be selling or discontinuing the former Highland Division services and did not believe a joint venture was now realistic despite accepting the economic logic. The years of discussions and negotiations with British Airways seemed to be going nowhere.

In December 1993, the Board now came to the conclusion that Loganair could not get out of the bit by itself. It was decided that an exercise be undertaken to consider the feasibility of Loganair merging with British Midland and to identify possible economies. A British Midland Express concept evolved and it was even believed that this could be up and running by 1 April 1994 with a separate Company to operate the combined scheduled services of Manx Airlines (Europe) and Loganair. Manx Airlines Ltd would stay outside this arrangement.

If the British Midland Express company was set up, Loganair would be subsumed and its identity lost. When it was first mooted, the author had argued that it would not be in the best interests of the Company's Highlands and Islands operations. He asked for permission to mount a Management Buy Out. Messrs Bishop, Wolfe, and Terry Liddiard, Managing Director of Manx Airlines, then gave much consideration to the 'rationalisation' of Loganair and the possible closure of the Glasgow hangar, and route transfers, before approval for the MBO to proceed was given in late March 1994.

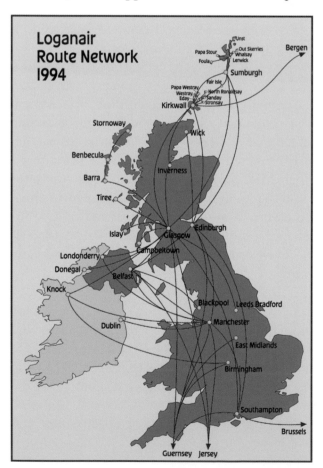

*The bigger you are the harder you fall! In 1994, Loganair's scheduled service network was at its most extensive and stretched from Unst, the most northerly point in Great Britain, to Jersey, the most southerly, and to Knock and Donegal in the west of Ireland. In 1993/94, some 660,000 passengers, Loganair's greatest number, were carried on scheduled services. Manchester was the major hub with links to Edinburgh, Belfast, Londonderry, Dublin, Knock, Southampton, Guernsey and Jersey.*

*On 31 March 1994, all Loganair's BAe ATP and Jetstream 41 services from the Company's Glasgow, Edinburgh, Belfast and Manchester bases, outwith the Scottish Highlands and Islands, transferred to Manx Airlines (Europe). 134 staff were made redundant and 268 Loganair staff were transferred to Manx employment.*

Meanwhile, the British Midland Express concept did not see the light of day. Instead, all eyes turned to the Isle of Man and to Manx Airlines. The Board took a dramatic decision. All Loganair's services outwith the Scottish Highlands and Islands, from its Glasgow, Edinburgh, Manchester and Belfast bases, which had been secured and developed with much blood, sweat and tears by Loganair's management over the previous twelve years, were now to be transferred to Manx Airlines (Europe) in the Isle of Man. Along with these routes, all Loganair's ATP and J41 aircraft would go to Manx. There would be some one hundred and thirty-four redundancies among Loganair's staff and two hundred and sixty-eight Loganair employees were to be transferred to Manx Airlines (Europe) on 31 March 1994. Certainly the saddest day in the Company's history.

Not surprisingly, when this staff announcement was made at the end of January, Company morale plummeted. There was one incident which made unfortunate headlines in the national press when a disaffected employee in Loganair's Glasgow engineering staff applied superglue to the flap control levers thereby immobilising an aircraft in the hangar. These were bad times.

Once approval was given for the MBO, the author invited five colleagues to join him. Much effort and professional fees went into preparing the Business Plan necessary for raising the funding package. The price required by the owners was agreed and the financial support secured from the 3i Group plc and the Bank of Scotland. All was now in place. To the great disappointment of the MBO team, and to the intense irritation of the two financial institutions, Sir Michael Bishop would prevaricate for some months. In July, the question of Airlines of Britain Group retaining twenty-five percent of the new Company with significant representation on the Board was raised but was not acceptable to the MBO team, nor to its institutional funders. How the different routes might be operated in future was also exercising the minds of the ABH Board. It was later in the summer before Sir Michael announced that Loganair would remain, after all, within the Airlines of Britain Group. He had changed his mind.

### Loganair as British Airways Franchise Partner

Ironically, almost at the same time, and after all these years, discussions with British Airways actually bore fruit and on 1 July 1994 Loganair became a British Airways Franchise Partner. It had become clear that the intense rivalry which existed between British Airways and British Midland, as well as Industrial Relations issues for British Airways, would dash any immediate prospects of a joint venture or a merger between BAHD and Loganair which had been the subject of very recent discussions between the two Companies. However, under the Franchise Agreement Loganair would operate, on behalf of British Airways, not only all its own Highlands and Islands services but also British Airways' Glasgow–Aberdeen, Glasgow–Inverness–Kirkwall–Sumburgh and Glasgow–Belfast routes, all with eight Shorts 360s and five Britten-Norman Islanders.

The joint British Airways/Loganair Franchise press statement emphasised that the new arrangement was to protect loss-making Scottish lifeline services and to stem further job losses. Not for the first time, an announcement by British Airways about changes in its Scottish operations was viewed with concern and suspicion. It was left to the author and British Airways' battle-hardened Scottish Manager, Ian Reid, to 'sell' the franchise concept and to travel round the Highlands and Islands reassuring Councils and the public that the franchise was good news.

*Courtesy of British Airways*

*Loganair became a British Airways Franchise Partner in 1994 and its scheduled services were marketed as connecting with British Airways' flights. Passengers could check in anywhere on British Airways' worldwide network and fly on a British Airways liveried aircraft to their final destination. In its sales and marketing brochures, Loganair proudly offered the British Airways passenger a 'seamless' transfer from one aircraft to another. The Concorde passenger might just have noticed the difference.*

Being General Manager of British Airways in Scotland during a period when British Airways was announcing changes or withdrawing from routes in, or to, Scotland, was hardly a sinecure. Never mind. The experience Ian Reid gained while explaining away the new Loganair franchise in 1994 would stand him in good stead when he had to face the public's wrath, after Robert Ayling reaffirmed British Airways' continued commitment to Scotland and then promptly announced the withdrawal of its Glasgow–Boston-New York service. Again it was Ian Reid who had to endure the public opprobrium when British Airways withdrew its Inverness–London Heathrow service, and it would be Ian Reid who had to face the Councils when British Airways suddenly announced it was giving up its six services to the Western Isles and Orkney and Shetland in 1996.

The author's task was a little different. He had somehow to get across to a genuinely confused public that, as a British Airways franchise partner, Loganair aircraft would in future be painted in a British Airways livery and Loganair's cabin staff's tartan uniform would be replaced by a British Airways uniform. Loganair would continue to be a quite separate company from British Airways. Yes, he could confirm that Loganair was still owned by Sir Michael Bishop's ABH Group and in the same stable as British Midland – and yes, indeed, it was the same British Midland that was British Airways' main competitor. A Henry Kissinger was needed.

### *A new British Airways timetable – 'Subject to Tides'*
Hardly had the ink dried on the British Airways Franchise Agreement when there was 'a little local difficulty' in Barra which caused very public embarrassment to both British Airways and Loganair alike. For the new franchise network, Loganair's fleet would be Shorts 360s and Britten-Norman Islanders. No Twin Otters were mentioned although the Glasgow Barra service with its cockleshell beach airstrip operations had for many years used a Twin Otter. Trials had been carried out on Shorts 360 operations on the beach, apparently to the satisfaction of the CAA. This was a welcome development for the Company as it allowed the last remaining Twin Otter to be sold, providing much needed cash to ease the Company's severe difficulties with the Bank. The Company also believed that the Shorts 360 signified an upgraded service with improved passenger comfort and increased passenger and freight capacity.

*In 1994, to provide more passenger capacity and to allow sale of the Company's last Twin Otter, the Company controversially introduced the Shorts 360 to beach operations for several months before public clamour demanded the Twin Otter be reinstated.*

Courtesy of Loganair

The Shorts 360 duly entered service, but the CAA imposed such severe operational restrictions one after the other, especially new rules relating to standing water on the beach, that these caused unacceptable service disruption and unreliability. There was much criticism from the local community accompanied by a high profile campaign to have the Twin Otter restored. Several television and radio companies, and many national newspapers covered the story all summer. Calum MacDonald, MP for the Western Isles, took up the cudgels and tabled a Parliamentary Question. Lord James Douglas-Hamilton, Minister at the Scottish Office also got involved. After eight months of service disruption and delays, and a great deal of local hostility, Loganair capitulated. It resisted serious temptation to give statutory notice of contract termination, but instead continued its commitment to Barra air services. A Twin Otter was sourced from Widerøe in Norway and again added to the fleet.

After the political furore, peace broke out again in Barra with the reintroduction of the Twin Otter. Later, local Councillor Donald Manford commented at the Company's twenty-fifth anniversary celebration of the Barra service that this was not a time for politics. He recalled the event which inspired the film *Whisky Galore* and commented

*The cockle strand at Tràigh Mhòr on the Isle of Barra is as firm as the typical HIAL runway, and aircraft have been landing there since before 1936 when Captain David Barclay inaugurated services from Glasgow for Northern & Scottish Airways. Loganair took over services to Barra from British Airways in 1975.*

Photo: Iain Hutchison

*The runway, of course, is washed twice a day! Even Loganair's Twin Otter, with its fixed undercarriage, requires additional washing after a beach landing to remove the highly corrosive salt and sand.*

*Following the closure of de Havilland's Twin Otter production line in 1987, Loganair's increasing difficulties in sourcing spare parts, and deterioration of the beach due to shell and sand extraction, there was concern about the future of the air service. From 1992, the Company collaborated with HIAL to identify possible sites on Barra and Vatersay for a fixed runway. A site for a 1,200 metre runway, just north of the present Terminal Building, was found. Concern at the possible £14 million cost for what would be a limited operation, and the re-start of Twin Otter manufacture in 2008 by Viking Aviation in Victoria, British Columbia, combined to ensure that beach operations will continue.*

that 'the only *Politician* that ever brought us any pleasure lies in ten fathoms off Eriskay.'

The episode, however, had highlighted the vulnerability of the beach operation and there was a referendum undertaken in 1996 on the construction of a hard runway, probably on the Eoligarry peninsula. This, however, produced an ambiguous outcome. In any event, the significant cost and severe operational limitations of a hard runway would probably ensure that Tràigh Mhòr was used as Barra's airstrip for many years to come.

### More Changes at Loganair. A Management Buy Out? Again

While Sir Michael Bishop was taking his time in the summer of 1994 to consider and then reject the MBO, Terry Liddiard, the Manx Managing Director was proposing a new structure for the Group's three regional airlines. A new Holding Company, Manx Airlines Holdings, itself a wholly owned subsidiary of Airlines of Britain Holdings, would assume responsibility for Loganair, as well as Manx Airlines and Manx Airlines (Europe). Terry Liddiard would become Managing Director and the author would remain Managing Director of Loganair and deputy to Terry Liddiard on the Holding Company Board. It was also agreed that Terry Liddiard and Captain Norman Brewitt, Manx's Director of Flight Operations, would be appointed to the Loganair Board, and that the quarterly Board Meetings of the three regional airlines, including Loganair, would in future be held in the Isle of Man. To confuse the travelling public even more, Manx Airlines (Europe) followed Loganair's lead and became a British Airways Franchise Partner.

Despite only a few months having passed since Loganair became a British Airways Franchise Partner, and Scotland just getting used to the new arrangements, John Wolfe and Terry Liddiard were busy driving forward their plan to have Loganair fully absorbed with all functions transferred to the Isle of Man. The author protested to Michael Bishop. The heart and soul would be ripped out of the Company, and where was the sense of creating further redundancies in Scotland just to take on additional staff in the Isle of Man? Only a few months earlier, Loganair's staff numbers had been reduced from more than six hundred to less than two hundred.

Despite John Wolfe's single-mindedness, and Terry Liddiard's determination to achieve full integration very quickly, it was still the author's fervent wish that Loganair remain as a Scottish-based separate legal entity, and it was fortunate that there were still some residual doubts and political sensitivities. After all, it was only a few months since much had been made in the press to emphasise that Loganair's British Airways Franchise was safeguarding the many lifeline services in the Scottish Highlands and Islands and many local jobs. The Scottish 'dimension' required careful handling, but also important to ABH, Loganair's £1.5 million bank overdraft facility with Royal Bank of Scotland might be put in jeopardy. Sir Michael Bishop invited the author to consider a Management Buy Out for a second time.

The author dusted down the MBO papers from the previous year, amended and updated them in view of the new British Airways Franchise. He again secured the financial backing of the 3i Group plc and the Bank of Scotland, despite both of them having been thoroughly disenchanted the previous year at Michael Bishop's last minute change of heart. The 3i Group plc was prepared to underwrite the full institutional equity investment, and the finance was raised by combination of preference shares, and bank loans and facilities, provided by the Bank of Scotland. The Loganair MBO team was all set. History, however, was to repeat itself and at each of the following two quarterly Board Meetings, Michael Bishop reported that no decision had been taken and it was not until December 1995 that he announced that Loganair would remain within the Group. For a second time, the author and his colleagues were frustrated, and two Scottish Financial Institutions supporting the Loganair bid were totally perplexed.

Courtesy of Loganair

*Loganair's Shorts 360 also received more than its fair share of comment. On one occasion, a lady passenger arriving at Glasgow Airport on British Airways' flight from Boston, and transferring onto Loganair's Shorts 360 flight to Islay, apparently got the fright of her life. She thought she was still on the airport bus, when the aircraft suddenly took off.*

***Dramatic developments certainly, but still a few loose ends.***

On 13 December 1995, Manx Airlines (Europe) announced 'another major expansion' once its management of Loganair operations had taken place. Manx Airlines (Europe) had also become a British Airways Franchise Partner like Loganair, and the two British Airways Franchises were now to be combined. It seemed there would be nothing to stop Loganair being absorbed into the one operating company by April 1996 and simply disappearing.

The prospect of being British Airways' biggest Franchise certainly appealed to Terry Liddiard, but there were still a few loose ends to be tied up first. The Public Service Obligation Contracts (PSOs) in the Northern Isles with the Orkney and Shetland Islands Councils were in the name of Loganair Ltd. It was the Company also which held the contracts with the Scottish Executive for the provision of air services from Glasgow to Tiree and Barra. These were lifeline services and politically sensitive. In any event, it was becoming clear that there was no great appetite in the Isle of Man for their Grand Plan to include eight-seat Britten Norman Islander and eighteen-seat de Havilland Twin Otter operations.

Then, in August 1996, there was the dramatic announcement. British Airways would be withdrawing from 'six loss-making services' carrying 250,000 passengers a year to the Western Isles, Orkney and Shetland, and Manx Airlines (Europe), flying under the British Airways Express franchise banner, would be taking over. The routes in question were Glasgow–Stornoway, Glasgow–Benbecula, Inverness–Stornoway, Aberdeen–Kirkwall, Aberdeen–Sumburgh and Kirkwall–Sumburgh. British Airways was continuing to lose money, was anxious to rid itself of considerable cost and had come to the decision that this could be achieved, without losing feeder traffic for its Domestic and International services, if the BAHD routes were handed over to a Franchise Partner. These routes and services were to be handed over to Manx Airlines (Europe) which would change its name to British Regional Airlines Ltd (BRAL) on 1 September 1996.

British Airways' announcement that it was pulling out of Scotland was not exactly welcomed either by the communities nor the politicians in the Northern Isles or Western Isles of Scotland. There was uproar. In particular, Calum MacDonald, MP for the Western Isles, was appalled at the way this had been done, and he and Jim Wallace, MP for Orkney and Shetland, wrote to Sir Colin Marshall urging British Airways to honour its commitment to the Isles. The Conveners of the Western Isles and Orkney Islands Councils, Alex MacDonald and Hugh Halcro-Johnston also weighed in demanding reassurances. The Secretary of State for Scotland, Michael Forsyth, then got involved for good measure.

With all this going on, Terry Liddiard was fearful that British Airways might change its decision in the face of such hostile opposition and criticism. He was aware that public opinion in Scotland could be orchestrated and effective, especially where transport was involved. Vice-Convener Angus Graham of Comhairle nan Eilean Siar, for example, was still flushed with his apparent success in having the decision to withdraw the London–Inverness night sleeper service reversed just a few months earlier. He now wrote directly to British Airways' Chief Executive, Robert Ayling, urging him to reconsider his decision and questioning whether 'a small regional airline company based in the Isle of Man can provide the same standard of service by merely changing the paintwork on the aircraft.' Now, such was its strength of feeling about this latest bombshell, that the Council had actually purchased a share in British Airways to allow attendance at the British Airways Annual General Meeting.

Scottish sensitivities were having to be taken seriously and the problem with Loganair's PSO Contracts had not gone away. Terry Liddiard was having a difficult enough time absorbing the British Airways routes. There was bound to be further political outcry if the Twin Otter and Islander operations were transferred to the Isle of Man and Loganair was closed down, as planned. It was probably more trouble than it was worth. After all, Scott Grier was still there. Sir Michael Bishop would surely be agreeable to a Management Buy Out this time. Or would he?

### The Third MBO Attempt

A feasibility study was immediately carried out on what was left of Loganair's flying operation. Gone were the Shorts 360 aircraft, and of course long gone were the BAe ATPs and the Jetstream 41 and 31 aircraft. Concentration was now on the earning potential of five Britten-Norman Islanders and a single de Havilland Twin Otter. The Islanders would undertake the inter-island air services in Orkney and Shetland, the Kirkwall–Wick scheduled service and, crucially, the Scottish Ambulance Contract which required Islander aircraft at Glasgow, Kirkwall and Lerwick. The single Twin Otter would undertake the Glasgow to Tiree and Barra services and the Barra–Benbecula service.

In addition to the flying operations, there would be a contract with BRAL for customer services to be provided by Loganair's existing staff at the outstations and also engineering support at Kirkwall and Stornoway. There would also be a contract to supply fuel at Benbecula for the MOD. It was apparent that the business case based on this very limited operation was always going to be very marginal despite all the flying activities, with the exception of the Kirkwall–Wick scheduled service, being supported by different Government contracts.

There were other very marginal or problematic Scottish services remaining in BRAL's network that were a source of great interest to the Loganair team. The

Glasgow–Campbeltown service, for example, had seen a serious decline in traffic since the MOD reduced the service complement at RAF Machrihanish in 1992 and subsequently placing it on a 'care and maintenance' basis in 1994. Servicemen had been regular users of the service. Now the main users were hospital outpatients, business users and golfers. Despite the acknowledged importance of the air service to the economy of the Mull of Kintyre and to local businesses like Jaeger, Campbeltown Creamery and Springbank Distillery, which provided vital jobs, Strathclyde Regional Council had consistently refused to subsidise the service. The Scottish Executive, too, faced difficulties with subsidy due to the new European Public Service Obligation rules. By the time of the MBO, the future of the service was in considerable doubt. It might take time, but if ever a PSO were awarded to the route, 'new' Loganair would be most interested.

Similarly, in the Western Isles, the PSO route Stornoway–Benbecula–Barra was in deficit, and air passenger traffic was plummeting on the 36-seat Shorts 360 Stornoway–Benbecula leg following the recent introduction of the new Sound of Harris vehicular ferry. The Benbecula–Barra sector, because of the beach landing, had to be operated by Twin Otter. If the MBO went ahead, BRAL would require to wet lease this aircraft from Loganair. The new Company business plan would clearly benefit by including this operation and at the same time BRAL would be rid of an operationally difficult and loss-making service. Terry Liddiard agreed to this, despite the Long Island service having being integrated with the more worthwhile Royal Mail Contract. Since 1993, the daily task of the Shorts 360 used on the Stornoway–Benbecula service involved the aircraft flying each morning with mail from Glasgow to Benbecula and Stornoway where the aircraft then undertook the two scheduled services between Stornoway and Benbecula before returning with mail to Glasgow. Clearly this would have been helpful to the 'new' Loganair and also to BRAL. Sir Michael would hardly object since including the Royal Mail contract inflated the MBO price for the Company.

This time the difficulties did not come from the Isle of Man nor Donington Hall. The Western Isles Islands Council, now officially renamed Comhairle nan Eilean Siar (CnES), firstly was adamant that no additional subsidy would be forthcoming despite the obvious impact of the new ferry on air passenger traffic. It was also most reluctant to allow its contract 'currently with a member of the British Regional Airlines Group with the financial stability, and the availability of resources that affords, to be transferred to a totally new Company with uncertain assets and no previous trading record.' 'People in faraway places are playing ping pong with our air services,' was Vice Convener Angus Graham's verdict, while Calum MacDonald MP favoured 'the musical chairs' analogy.

However, it was the CAA in the end which created the greatest difficulty. Operating the Shorts 360 required a Type A licence while only a Type B licence was required for the Islander and the Twin Otter. The necessary formalities and paperwork could not be completed in time for the start of the new Company. The Stornoway–Benbecula–Barra service and the Royal Mail Contract sadly had to be withdrawn from the new Company's business plan. Still, in the minds of the Loganair's management, this could be a real future prospect.

The author would insist on a number of preconditions. The new Company would be called Loganair as he believed that it was held in high esteem and affection in many parts of the Highlands and Islands because of its long history and the commitment it had demonstrated for more than thirty years. Nor would the Management Buy Out proceed without the 'new' Loganair having a guarantee that the British Airways Franchise would continue. 'Old' Loganair had been a Franchise Partner since 1994. The new Company continuing with British Airways would mean the aircraft flying in British Airways livery, staff would be in British Airways uniform, and this would give a credibility and clear signal to the travelling public that Loganair was conforming to British Airways' high operating standards. Moreover, being part of British Airways' worldwide scheduled service network, having the benefit of its sales and distribution systems, and having British Airways provide reservations and revenue accounting functions, would also be of immense value to the fledgling airline. This was vital. The author was grateful to Lewis Scard, the British Airways Franchise Manager, who in a huge act of faith persuaded his Board to grant a six-year Franchise to Loganair following a satisfactory British Airways audit of all Loganair's operational and maintenance procedures.

There were many other issues to be resolved. Nothing was straightforward. In Shetland, BRAL's MOD contract to transfer servicemen between Edinburgh and RAF Saxa Vord on Unst clearly required re-negotiation as the Sumburgh–Unst element was operated by Loganair's Britten-Norman Islander. Even more problematical was the proposed customer services contract which was intended to add much needed substance to the 'new' Loganair. It had been assumed that the British Airways staff at Sumburgh, Kirkwall, Stornoway and Benbecula airports would transfer to Loganair to allow the new Company to provide customer services for BRAL at these airports. British Airways prevaricated and it was to be another six years before its staff would transfer to Loganair employment. Meanwhile, the new Company had to be content with a much smaller contract to provide grounds services for BRAL at airports where Loganair's own staff were employed.

Terry Liddiard and his colleagues in the Isle of Man meanwhile were working tirelessly to make operational and commercial sense of BRAL's suddenly enlarged network. The MBO would not be permitted to be a distraction and they wanted the MBO to take place sooner rather than later. Nevertheless, it took a formal request from Terry Liddiard, supported by Austin Reid, before Sir Michael Bishop agreed to the MBO. 1 January 1997 was the target date, but there were still purchase price issues to be addressed.

Back in 1987, the British Midland Group of Companies had been reorganised: a new Holding Company, Airlines of Britain Holdings Limited, was set up and the minority shareholdings of Scott Grier and Air UK in Loganair and Manx Airlines respectively were bought out. Consequently, British Midland, Manx Airlines and Loganair became wholly owned subsidiaries of the new Airlines of Britain Holdings. Scandinavian Airline System (SAS) at the same time acquired 24.9 percent of ABH. Grier became a shareholder in ABH also, and these shares would be used to meet the purchase price of new Loganair. The balance would be met by Stephen Bond of Bond Helicopters Limited who was invited by the author to take a thirty percent shareholding in the new Company.

There would be a few more twists and turns, and not a few surprises and delays, before the deal was done – to the extent that there was a real concern that Sir Michael Bishop would change his mind a third time. Not that he was visible during the process. There had been no dialogue between him and the author for many months. He had delegated this to Austin Reid, the ABH Chief Executive, to progress matters. The first delay, rather bizarrely, seemed to be due to Sir Michael taking offence when Trevor Bush, who had previously left Loganair to join British Midland at East Midlands Airport, accepted the author's invitation to leave British Midland to come back to Loganair as Managing Director.

Austin Reid managed to get the process back on track. Then, at the eleventh hour, he conveyed the potentially disastrous message, without explanation, that there had been a sudden reduction in the value of the ABH shares which the author held and which were being used to meet the purchase price.

Then, perhaps most strangely of all, there was a further delay at the eleventh hour when Sir Michael Bishop instructed Hugh O'Donovan of law firm Wilde Sapte to look afresh at the Sale Agreement. The threat to British Midland imposed by Loganair's five Islanders and single Twin Otter was not immediately apparent to many observers. The author professed himself to be flattered.

Eventually, on 28 February 1997, the Purchase Agreement was duly signed. There was no triumphalism, only a sense of overwhelming relief. From the first abortive MBO in 1994 it had been a long rocky journey, often frustrating and sometimes heartbreaking. Clearly there were challenges ahead, but the management took great comfort in the knowledge that the extremely difficult relationship with Sir Michael was now at an end, and the new Chairman had rather more empathy with what Loganair was trying to do in Scotland. The next episode in the history of Loganair was about to begin.

# Chapter 4

# Scottish Again
# 1997 - 2012

### The 'New' Loganair Business Plan

At long last, on Saturday 1 March 1997, the new independent Loganair was up and running. For the sixty-nine full-time and seven part-time staff, there was relief that the deal with Sir Michael Bishop had been completed, but not surprisingly the management realised only too well the sheer scale of the challenge ahead. The Business Plan was sound, but the Company had been left with few aircraft and a tiny flying operation, and for the new Company to stay viable, new business would have to be obtained very soon.

Meanwhile the Business Plan's flying operation was based on a single, eighteen-seat de Havilland Twin Otter and five, eight-seat Britten-Norman Islanders. The Twin Otter would operate the Scottish Executive's Public Service Obligation (PSO) contract for provision of air services from Glasgow to Tiree and Barra. The same aircraft was also subcontracted by British Regional Airlines Ltd (BRAL) to operate the Benbecula–Barra scheduled service as part of its CnES PSO contract for the Stornoway–Benbecula–Barra route.

Four of the five BN2 Islanders were fully tasked while the fifth was designated to provide cover during maintenance periods. A dedicated BN2 aircraft was based in Glasgow to provide twenty-four hour cover for the Scottish Ambulance Service. Two Islanders were based at Kirkwall, one for the twenty-four-hour air ambulance services and the other for Orkney's North Isles passenger services and the Kirkwall–Wick scheduled service. One BN2 Islander was based in Lerwick, not only to provide inter-island passenger services for Shetland Islands Council and air ambulance services for the Scottish Ambulance Service, but also for lease to BRAL on the Sumburgh–Unst leg of its MOD contract to transfer personnel from Edinburgh to Unst via Sumburgh.

All the scheduled services had the benefit of being part of the British Airways Franchise Agreement which was to run for six years through to 2003. This included the Orkney and Shetland internal passenger services but later these would be excluded by British Airways at the Franchise renewal in 2003 because they were single pilot operations.

There were important non-flying activities included in the Business Plan, notably passenger handling services at the different Highlands and Islands Airports, and the

Fuelling Facility at Benbecula. Loganair's staff at Wick, Campbeltown and Islay provided third party passenger handling services for BRAL, and from 1 October 1997 also at Kirkwall, Sumburgh, Stornoway and Benbecula airports when the British Airways passenger handling staff at these airports would transfer to Loganair. At Benbecula, the fuel farm allowed the Company to undertake a two-year MOD Contract for the sale of kerosene - 511,000 litres per annum based on BP's fuel price plus an agreed premium. Again, to boost the income, the Company's engineering staff at Kirkwall provided maintenance checks and support for BRAL's Shorts 360 aircraft which was parked overnight in Orkney. Islander and Twin Otter pilot training was also provided for third parties.

The new airline's operations were clearly very small and fragile. The author, however, was not setting off altogether on a wing and a prayer. From the business perspective, the new operating model had a distinct advantage of being mostly underwritten by guaranteed contractual income– from the Scottish Office, the three Islands Councils, the Scottish Ambulance Service, BRAL and the MOD. Of all the Company's flying activities, only the Kirkwall–Wick scheduled service and four of the Orkney Island services were operated at Loganair's own commercial risk.

Much more importantly, the Loganair management was firmly of the view that several of Loganair's former scheduled services which had been transferred en bloc to Manx/BRAL in 1994 and 1995 would not sit easily in BRAL's Grand Plan and might well become available again to the Company sooner rather than later. Terry Liddiard, BRAL's Chief Executive, after all, had said publicly on several occasions, usually in response to some stinging criticism of BRAL's services in the Highlands and Islands, that he would give the routes two or three years to prove themselves. At least he had assured the Secretary of State for Scotland on 4 December 1996 that if any change were to be made, advance warning would be given. At that time, the MBO process was faltering, with one problem after another, and seemingly endless delays with no certainty that Sir Michael Bishop would not change his mind for a third time. Terry Liddiard's comments were just the kind of encouragement the author needed in order to stick with it.

### Under Way

The new Company was given an early boost. The Scottish Executive had eventually imposed a Public Service Obligation on the Glasgow–Campbeltown route, the same route which Strathclyde Regional Council had resolutely refused to subsidise in the early 1980s, and Loganair was awarded the contract for three years from 1 May 1997.

Being aviation, not everything went strictly according to plan. In Orkney, the Kirkwall–Wick scheduled service was further weakened by recent additional ferry services across the Pentland Firth. Of the inter-island air services, only the two air services to North Ronaldsay and Papa Westray which had no roll-on roll-off ferry services, were subsidised by the Council and the other four North Isles services to Sanday, Stronsay, Eday and Westray, were losing money. Negotiations for subsidy began with the Council, but it was 2002 before all six North Isles services were included in a PSO contract.

The single BN2 Islander operation based at Tingwall Airport in Shetland did go according to plan as the whole operation was effectively underwritten by the Council, the Scottish Ambulance Service and BRAL's MOD Contract. Overall, the main activities of the Company's Business Plan performed as expected. A small profit of £60,000 was achieved on a turnover of £4.4 million in the new Company's first financial period to 31 March 1998. The Loganair management took great satisfaction from this modest achievement, believing it might have been the first profit by an airline operating wholly within Scotland since the 1930s.

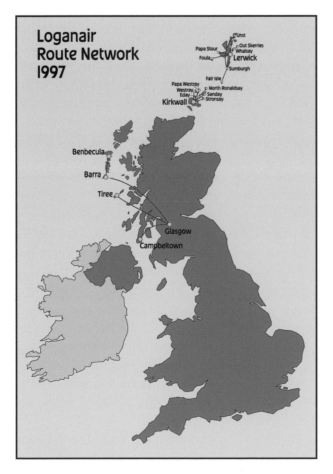

*At the beginning of the new Company, five Britten-Norman Islanders operated on the Shetland and Orkney inter-island air services while the de Havilland Twin Otter undertook the Scottish Executive contract for provision of air services from Glasgow to Tiree and to Barra. In May 1997, two months after the new Company started, the Twin Otter was also engaged on the new Glasgow–Campbeltown Public Service Obligation contract.*
*For the year ended March 1998, a profit of £90,800 was achieved on a turnover of £4.4 million with passenger carryings of 47,872. The aircraft fleet still comprised five Britten-Norman Islanders and one de Havilland Twin Otter.*

Even more encouraging for Loganair management, however, was the dialogue taking place with BRAL which would lead to Loganair taking back from them some of its former scheduled services in Scotland, no doubt the result of BRAL's decision to discontinue its Shorts 360 operations. The first two, the Glasgow–Islay and Glasgow-Londonderry routes, were taken over by Loganair on 1 January 1998. This was at very short notice, and Terry Liddiard agreed to lease a Shorts 360 with crews to Loganair while the Company acquired an aircraft and recruited and trained its own pilots. This was the early breakthrough the new Company needed. Other routes would surely follow and during the second year of the new Company, in November 1998, the Glasgow–Inverness–Kirkwall route and the Edinburgh–Wick–Sumburgh route were transferred to the Company.

Good fortune for Loganair, but BRAL's decision to withdraw was very badly received in the Northern Isles. As recently as February 1998, BRAL had announced that the unpressurised Shorts 360 were being replaced by Jetstream 41 aircraft on these routes: 'The Shorts 360s have run out of life and they are not comfortable really. The Jetstream 41s will reduce flight time considerably and are a lot more comfortable.' For Orkney Islands Council Convener, Hugh Halcro-Johnston, Loganair coming back on the route again meant Shorts 360s and this was 'a matter of grave concern'. Any significant improvement on the route now looked 'unlikely'. This adverse public reaction merely confirmed the view of Loganair's management that, for the Company to gain acceptance as rightful successor to British Airways on the Scottish routes, a pressurised aircraft was required. The thirty-four seat Saab 340B aircraft was identified. The Company's Shorts 360 services would be operated for as short a period as possible until the Saabs could be acquired and crews trained, hopefully in time for a summer 1999 start.

The Company was also awarded 'the Long Island' route, the Stornoway–Benbecula–Barra service, by the Council when BRAL, which had taken it over earlier from Loganair, decided not to tender. At the same time, the Company submitted a bid for the Royal Mail Contract *Skynet 12* for the carriage of mail from Glasgow to Benbecula and Stornoway. This was a neat operation. At Stornoway, Loganair's engineers would fit the thirty-six passenger seats for the two daily passenger service rotations between Stornoway and Benbecula. On its return to Stornoway in late afternoon, the seats would be removed, the mail loaded and the aircraft would return to Glasgow. Neat, that is, except for the occasion when the seats were inadvertently stranded at the wrong end of the route. There was a very public announcement that the scheduled service to Benbecula had been cancelled, and why! The national press, not surprisingly, had a field day and there were red faces all round.

Being awarded another seven-year contract by Scottish Ambulance Service rounded off a very satisfactory second year's operations. The annual profit increased to £227,000 on a turnover of £9.6 million. The business was sound, and the Company was well and truly up and running. It had all happened very quickly. The routes taken back from BRAL were just what Loganair needed and, indeed, what it had been hoping for. The young Company's optimism and expansion continued the following year when it took over the Inverness-Stornoway route from British Airways in January 2000. It also had the confidence to start new, non-stop services on the Edinburgh-Stornoway and Edinburgh-Sumburgh routes. Unfortunately, these opportunities came along far too soon. The Company did not yet have the necessary resources and infrastructure. It was not coping and in financial year 1999/2000 it lost £402,000. This was Year 3 for the new Company - the most vulnerable for a start-up airline.

*To gain acceptance by the travelling public, 'new' Loganair recognised the need to replace the unpressurised Shorts 360 aircraft, and in 2000 took delivery of its first two thirty-four seat pressurised Saab 340B aircraft. Later, in 2005, it proved ideal for replacing the larger BAe ATP aircraft when the Company took over the last of British Airways' Highlands and Islands routes.*

Courtesy of Loganair

### The Board

Sir Michael Bishop had agreed to the author leading a Management Buy-out in 1994, only for him to change his mind at the eleventh hour. One year later, he had actually suggested that the author should consider once again an MBO, and once again he changed his mind and retained a much emasculated Loganair in the Airlines of Britain Holdings Group. For both of these abortive MBO attempts, funding arrangements were in place through 3i plc and the Bank of Scotland. By the time the author was permitted to bid again for Loganair in 1996/97, it was but a rump of the earlier Company. This time, no institutional finance was required. The purchase price, in the main, was met by exchanging the author's shareholding in ABH. The author offered a minority shareholding to Stephen Bond whose company, Bond Aviation Services, since 1993, had provided the helicopter service element of the Scottish Ambulance Contract, which of course was also one of Loganair's biggest contracts. The author had initially expected Bond Aviation Services to be the shareholder, but Stephen Bond had sold his company in the interim to a Norwegian helicopter company and he agreed to purchase the shareholding himself, and to become a non-executive director of the new Loganair. In 2007, Stephen Bond increased his shareholding to 50% less one share.

On 1 March 1997, the Board of the new Company therefore comprised Scott Grier, Chairman, and Stephen Bond as a non-Executive Director. Trevor Bush, who had headed up Loganair's engineering department since 1988 and who had been Scott Grier's Deputy in the old Company until he was transferred to British Midland in 1995, now returned to Scotland and to Loganair as Managing Director. He guided the Company through the early years until he retired in September 2000. David Ross, who had been Loganair's solicitor since 1985, became Company Secretary. When he retired from his firm, Biggart Baillie, in 2009, he became a non-Executive Director. He was joining another non-executive director, Lloyd Cromwell Griffiths, who was appointed to the Loganair Board in 2008 on his retiral as British Airways' Director of Flight Operations and he became the chairman of the Company's Flight Safety Review Board.

In 2001, Jim Cameron, formerly of British Airways and Brymon Airways, succeeded Trevor Bush as Managing Director and was in post until 2006. Peter Tierney was recruited from Vancouver, Canada, to take over as Chief Executive, but after only a year in office his place was taken by David Harrison, the Company's former Director of Finance. David Harrison formed a strong management team that included Jonathan Hinkles, who had previously been Managing Director of Zoom Airlines, as Loganair's Commercial Director.

### Highlands and Islands Airports

In aviation terms Scotland is but a small village. How Loganair has performed over the years has always been influenced by other airline activity or competition in the area. For decades, British Airways was all-powerful and protected by the CAA licensing regime. For long periods, Loganair's destiny was largely in the hands of British Airways and what crumbs it could glean from the master's table. There were other players, of course, which were trying to establish a niche for themselves in a very tight Scottish environment. Air Ecosse, for a decade from the late 1970s and, more recently, Highland Airways both tried and failed to survive in this difficult marketplace.

There was another non-airline player, however, which was always in a position to help or to hinder Loganair's progress. Highlands and Islands Airports Ltd (HIAL) had been incorporated in 1984 as a wholly owned subsidiary of the CAA to take over the running of its eight Highlands and Islands airfields – Islay, Tiree, Benbecula, Stornoway, Inverness, Wick, Kirkwall and Sumburgh.

The previous year, there had been serious concern throughout the Highlands and Islands when Iain Sproat, the Aviation Minister, had announced in the House of Commons the Government's intention to sell off the Highlands and Islands aerodromes as part of Mrs Thatcher's privatisation programme. This, not surprisingly, proved to be no easy task in view of the aerodromes' history of financial losses. Only Sumburgh Airport was profitable, having had the £13 million debt relating to the building of the Wilsness Terminal written off. Even Sumburgh, however, was problematical as, all too quickly, the airport traffic was dwindling as more and more oil support flying had switched to Aberdeen with the advent of the longer range Chinook helicopters.

To the disappointment of the Shetland Islands Council, which confirmed its interest in taking over Sumburgh Airport, the Department of Trade confirmed that local authorities were ruled out because buyers, whether individuals, companies or perhaps consortia, were to be sought from the private sector. Loganair briefly considered taking over Islay and Tiree aerodromes where it was the sole operator, and it was understood that Air Ecosse had considered, albeit briefly, the possibility of acquiring Wick. In the event, there were no credible takers and HIAL was formed to take over the running of the eight airports. Later, in 1994, HIAL took over Barra aerodrome from Loganair when the airfield classification was changed at the time of the very brief introduction of Loganair's Shorts 360. In the same year, the ownership of HIAL was transferred from the CAA to the Secretary of State for Scotland (and later the Scottish Government). In 1996, HIAL received a dowry of £1 million to take over RAF Machrihanish (Campbeltown) from the MOD, and more recently, in 2007, it added Dundee Airport to its portfolio to make a total of eleven airports.

With the exception of Inverness Airport, for some years Loganair has been HIAL's biggest customer. Indeed at Campbeltown, Islay, Tiree, Barra and Benbecula, Loganair has been HIAL's only regular operator. For the same reasons as airlines have struggled historically in Scotland – small population catchment areas and passenger traffic, combined with expensive fixed costs and difficult operating environment – HIAL incurs financial losses at all eleven of its airports and is totally dependent upon subsidy from the Scottish Government.

Loganair always empathised with HIAL's Mission Statement which is 'to provide and operate safe, secure and efficient airports which support the communities we serve.' More than that, HIAL and Loganair are both very fortunate to have in common many staff members at the different airports who are wholly committed to serving their communities. Not surprisingly, therefore, Loganair has always seen its role and HIAL's role in the Scottish Highlands and Islands, as identical. Indeed, during the Chairmanship of Sandy Matheson and, more recently, of Grenville Johnston, HIAL and Loganair seemed to have common purpose and much progress has been made by constructive collaboration. Alas, it was not always so. In fact there were times during Peter Grant's chairmanship when Loganair was firmly of the view that customer-airlines and passengers were something of a nuisance to HIAL.

Airfield opening hours have been a cause célèbre since the early, post-nationalisation years of Scottish aviation. Naturally, Loganair, and indeed all airlines, have to schedule their services within HIAL's published airfield opening hours which were formerly very restricted and clearly a major obstacle to airlines achieving adequate aircraft utilisation. Airlines in the Scottish Highlands had to operate, and try to make a financial return, in a shorter period in the day than airlines elsewhere in the UK. For years, Loganair requested HIAL to extend its hours of business. As recently as 2001, Stornoway Airport, possibly seventh busiest airport in Scotland, was open only from 0800 hours to 1730 hours each weekday, and till 1500 hours on Saturdays. Only by considerable scheduling effort within this operating environment, Loganair managed to achieve 3.5 hours utilisation per aircraft per day. By comparison, elsewhere in the UK, British Midland was managing to fly its aircraft 8.5 hours each day, and EasyJet 10.2 hours per day. It was a struggle for the Company.

Even more reprehensible in Loganair's eyes was HIAL's policy of 'disincentive' pricing of out of hours airport charges, especially at weekends when many of the airports had even more limited opening hours, some closing from lunchtime on Saturday until Monday morning. Any airline worth its salt would try to fly its aircraft harder, but additional work outwith HIAL's published airfield hours was faced with punitive airport charges, based not on cost plus profit margin, which would have been perfectly reasonable, but at a level that were likely to discourage potential customers. It seemed HIAL simply did not want to open up its airports outwith normal hours for customers.

Then there was the issue of navigational aids, which were limited at HIAL's airports and this inevitably affected the airline's scheduled service reliability. One of Sumburgh's two runways had an Instrument Landing System (ILS), but at the other airports, including Kirkwall, there were non-precision approach aids and standard Airfield Ground Lighting. Matters came to a head at Kirkwall. The exceptionally bad weather in the summers of 1997 and 1998 had prompted a public clamour for an ILS for Kirkwall to help improve the airline's scheduled service regularity and reduce service disruption.

HIAL's Board showed reluctance to accept the merits of ILS, but commissioned a study, *Concept and Development of Approach Aids at Kirkwall Airport*. To the frustration and disbelief of all interested parties, the study chose 'not to quantify the social or business cost to the community' but concentrated on the effect of weather on HIAL's own finances. It concluded that as few as two percent of flights were cancelled, which deprived HIAL's coffers of a mere £17,500 a year in lost landing charges. Most parties campaigning for ILS realised it was never a commercial proposition for HIAL, but for the study to ignore the delays and disruption, the cost and inconvenience to passengers, and the interests of airlines and the community at large, was disappointing in the extreme. The Orkney Islands Council kept up the pressure and was supported vigorously by Member of Parliament, Jim Wallace, and by Loganair's Senior Pilot in Kirkwall, Stuart Linklater, whose knowledge of local flying weather and conditions was invaluable. When, eventually, ILS was installed at Kirkwall Airport, scheduled service reliability and overall performance improved dramatically.

For 'New' Loganair in the late 1990s, HIAL's apparent inflexibility and reluctance to review its practices, especially its airfield opening hours, were a source of great frustration. For a long time, it was not accepted by HIAL that its Air Traffic Controllers could exercise discretion on occasion with regard to duty hours. As a consequence, the Loganair aircraft and, of course, its passengers could be overhead Kirkwall Airport at airfield closing time, be refused permission to land and have to return to Inverness. Or be sitting at the end of the Kirkwall runway at closing time and not be permitted to take off, resulting in a night in a hotel for all the passengers. In early 2000, independent consultants were appointed to carry out a Pricing Strategy Review for HIAL and their report included the comment that 'HIAL did not respond in a friendly fashion to its customers' requirements.' HIAL's Chairman, Peter Grant, seemed to believe that this criticism emanated from Loganair.

The author was anxious to meet Mr Grant at this time as there were many long-outstanding matters to be discussed, including not least of all, Loganair's serious, but short-term, cash flow difficulties which had been detailed with explanations to Robert Macleod, HIAL's Managing Director. Despite repeated written requests for a meeting, Peter Grant could not find time to meet the author during an eight-month period before he retired as Chairman in February 2001. HIAL, however, did have time on 13 September 2000, through Managing Director, Robert Macleod, to threaten Loganair with legal action within twenty-four hours, which would have effectively put the Company out of business for having exceeded agreed credit terms. Not for the first time, Loganair scraped by. The author later drew HIAL's actions to the attention of Ms Sarah Boyack MSP, the Minister of Transport. He wrote: 'To do so in the middle of a fuel crisis showed a breathtaking insensitivity and callous disregard for the many remote communities which depend on Loganair's services.'

Loganair's pleasure and relief at the change of the HIAL regime on 1 March 2001 was vindicated when new Chairman, Sandy Matheson, was actually prepared to meet HIAL's customers and give consideration to suggestions or requests. HIAL's new 'Working In Partnership' approach was welcomed by airlines and communities alike. Gradually, the opening hours at the different airports were extended. Airlines now had the opportunity to provide additional services. Passenger traffic was stimulated and HIAL's revenue increased. The airlines benefited, the passengers benefited, and HIAL benefited – all worth waiting for, but it had been a long time coming.

In the following few years, HIAL worked with its airline partners and generally adopted a much more commercial approach. HIAL, understandably, wished to attract new business to its airports. Not everything would please Loganair, nor, it seems, Ryanair. HIAL Managing Director, Robert Macleod had a very public disagreement with Ryanair and probably received very unfair criticism from Michael O'Leary, Ryanair's Chief Executive, who accused HIAL of intransigence over landing charges at Stornoway airport, which he claimed had discouraged Ryanair from starting low-cost services to Stornoway.

There had been scheduled services for many years to Stornoway from Glasgow and Inverness, but only recently from Edinburgh. Loganair's Edinburgh–Stornoway passenger traffic was growing, but still thin. No doubt still smarting from Michael O'Leary's criticism, Robert Macleod clearly thought he had achieved something of a coup when British Midland agreed to start Embraer 145 jet services on the route. He chartered an aircraft to take journalists to Stornoway for a high profile media launch of the British Midland service which would be competing with Loganair. Most observers realised British Midland would be on the Stornoway route for only as long as its network timetable had a jet aircraft parked at Edinburgh Airport each day for just long enough to fit in a Stornoway return service. The only surprise, perhaps, was that British Midland stayed on the route for as long as it did – some three years, but not before much damage had been done to Loganair's own service. Loganair was strongly of the view that HIAL had overstepped the mark in so obviously supporting an expensive media launch for the services of one airline when another airline was already on the route. In the Company's view, it was a debatable use of taxpayers' money and was difficult to reconcile with HIAL's policy of even-handedness.

Loganair acknowledged that HIAL has a difficult role in trying to attract air services to its airports. In Scotland's small operating environment, developing a new air service almost invariably has a direct or indirect impact on neighbouring services.

When HIAL encouraged Eastern Airways to provide non-stop services from Inverness to Manchester and Birmingham, undoubtedly exactly what the traveller wanted, it was enough to end Loganair's Glasgow–Inverness service which had been dependent for many years on transfer passengers from Birmingham and Manchester at Glasgow. With Eastern's new non-stop services, these passengers now overflew Glasgow. For the first time since the Second World War, there was no scheduled air service between Glasgow and Inverness.

Loganair was grateful to HIAL in 2003 when, amidst some controversy, it agreed to open Stornoway Airport on Sundays. Until then, the Sabbatarian view had prevailed and there had never been Sunday air services to the Isle of Lewis. The author was received most courteously by the Lord's Day Observance Society (LDOS) when he explained that, in proposing Sunday services from Glasgow and Inverness, the Company was responding to mounting passenger demand. For its part, LDOS was fearful that public transport services on the Sabbath would offend a large proportion of the Lewis church-going population and change a way of life. Needless to say, the two parties agreed to disagree.

The author received some eleven hundred letters, mostly from people opposing the service, but also from many who made a valid argument for needing to fly on Sundays. The Company commenced services in September 2003 having given assurances that none of the Company's staff would be forced to work on Sundays if their religious beliefs forbade it. The Sunday scheduled services quickly grew in popularity and an Edinburgh–Stornoway service was soon added to the services from Glasgow and Inverness.

HIAL's role in Scottish aviation is huge. The efficiency, indeed the success of air services in the Highlands and Islands depend upon a constructive collaboration between HIAL and its airline customers. They have common purpose.

### *2001 – a year to forget*

For Her Majesty the Queen, it was 1992. For Loganair, 2001 was the *annus horribilis*. The year had started as the previous year had ended with the Company still having cash-flow difficulties. The previous September's experience of HIAL threatening the Company with closure was still a vivid memory. Trading continued to be difficult and was made worse by very severe weather and heavy snowfalls affecting operations at Glasgow and Edinburgh airports.

On 27 February, a Shorts 360 operating a night contract for Royal Mail on the Belfast–Edinburgh route had been unable to make the return leg due to snow at Edinburgh Airport. The aircraft remained on the tarmac for the rest of the night and throughout the following morning and afternoon. The weather improved sufficiently

Since the mid-1980s, Loganair has been offering golf packages on its scheduled services from Glasgow. Thousands of golfers have now enjoyed the delights of Machrihanish golf course, Machrie golf course on Islay, and Askernish on South Uist.

Machrihanish

Machrihanish Golf Club was founded in 1876. The course was described by 'Old' Tom Morris himself as 'Specifically designed by the Almighty for playing golf'. It has what many in the world believe to be the finest opening hole in golf.

Photo: John T Gillespie

Machrie

Laid out in 1881 by Willie Campbell, it has stunning panoramic views of the Mull of Oa and the Rhinns of Islay. It was the unlikely venue, in 1901, for a match to be played for £100 for the first time in the history of golf which was won by one of the 'Great Triumvirate', JH Taylor.

Photo: Getty Images

Askernish

The newly restored 'Old' Tom Morris course of 1891 was opened in 2008. Beautifully situated among the sand dunes, it has been hailed as 'The most natural golf course in the world'. In 1936, Northern & Scottish Airways started a regular service from Renfrew to Askernish landing on the flatter part of the golf course. After 1938, Askernish was used only as an 'on demand' stop and for landings by the Air Ambulance Service.

Photo: Getty Images

for the pilots, Captain Carl Mason and First Officer Russell Dixon, to undertake their pre-flight procedures including running the engines on the ground for some time prior to taxiing for take-off. A few minutes after take-off, both engines failed and both pilots died in the ensuing crash. The terrible double tragedy was deeply felt, not only by the pilots' families, but by everyone in what was very much a small close-knit airline. The Air Accident Investigation Branch subsequently recommended that de-icing veins on the aircraft's two engines should be switched on separately and not together, and engine intakes inspected and cleared of snow or ice if the aircraft was standing in extreme weather, and blanks covering the engine intakes should be deployed. Following the accident, Russell Dixon's fiancée presented the Carl Mason and Russell Dixon Memorial Trophy, which is awarded annually to the person in the Company who has made the greatest contribution to flight safety.

Trading continued to be difficult. Results from both the Orkney and Shetland inter-island services were disappointing, as was the return from the Scottish Ambulance Service contract. In the Western Isles, the contract for operating the Stornoway–Benbecula PSO contract now specified an eighteen-seat aircraft, instead of the thirty-six aircraft as formerly, due to the downturn in passenger demand following the introduction of the Sound of Harris ferry. The Company was not successful with its tender for the smaller aircraft operation and, worse still, Highland Airways, which was awarded the contract, was now in a position to introduce new competition on the Stornoway–Inverness route.

For some while after the accident there seemed to be a loss of public confidence in the Shorts 360 and some passenger resistance to using the Company's scheduled services. At the same time, heavy post-accident expenses were being incurred and a replacement aircraft required to be leased in. The Company's cash flow continued to be critical. To make matters much worse, significant costs were being incurred in introducing the new Saab 340 aircraft type which required significant pilot recruitment and training.

The Flight Operations management had acted professionally and effectively in relation to the Shorts 360 accident and its aftermath. There had been a number of other difficulties involving the Flight Operations management, notably some administration and training issues and difficulties in introducing the Saab 340A on to the Company's AOC. The Board decided that a management restructuring was appropriate and this prompted the resignation of three senior members of the Flight Operations Management, one of whom would withdraw his notice later and go on to play an important role in the Company for some while.

Morale was at an all time low and not improved by Loganair receiving notice of two Industrial Tribunals: from one of the senior Flight Operations Management team, and the other from a Shetland pilot who claimed that the Company had used the wrong mixture of disinfectant during the Foot and Mouth outbreak which had made him ill. Both cases would eventually fail, but not before bitter and regrettable recriminations had dominated management's attention all summer.

The 9/11 terrorist attack on the twin towers of the World Trade Center in New York on 11 September 2001 was the greatest crisis the aviation industry had faced since the oil crisis of the 1970s and had severe repercussions for the airline industry worldwide. Loganair was not immune. The immediate and most serious reaction was from the insurance industry, which had never before contemplated the possible, almost incalculable, liability from damage that aircraft could cause on the ground. For all airlines worldwide, third party liability insurance for war risk was withdrawn on seven days notice. Without the correct level of insurance in place, all UK aircraft were effectively grounded. Insurance cover, if it was obtainable, was prohibitively expensive. At the eleventh hour, as an interim measure, the UK Government made arrangements to allow aircraft to fly for thirty days.

Increased levels of security were imposed immediately at the major airports and new security arrangements introduced in the Western Isles, Campbeltown, Islay, Orkney and Shetland. Initially it was thought that the Government might pick up the bill for these enhanced security arrangements, but it became clear that these increased security costs were to be passed on to airlines and, ultimately, on to the passenger by way of surcharge. Only after much pleading were Loganair's Orkney and Shetland Inter-Island Islander services granted exemption as before. After 9/11, in common with the rest of the worldwide airline industry, the Company saw an immediate downturn in business, especially on the Company's many services with a high transfer-passenger content when the facility for through check-in passengers connecting with franchise partner British Airways was withdrawn. Fortunately these connections were restored after a short period.

### Real progress with scheduled services

Having been anxiously seeking new scheduled services to give some much needed substance to 'new' Loganair's business model, it was ironic that when the old Shorts 360 routes actually started to be surrendered by BRAL in 1998, the Company was ill-prepared for it and needed the help of BRAL's crews for a period. The following transition through Shorts 360 to Saab 340 aircraft had then been financially difficult and painful for the Company. Now, when in 2003, British Airways CitiExpress (BACX) handed over its last remaining Scottish routes to Loganair, the Company still had insufficient resources to react immediately. In fairness, this involved the highest volume routes in Scotland and was bound to be a major challenge to the Company.

British Airways CitiExpress was formed in 2001 when Brymon Airways, Manx,

Courtesy of Loganair

*Fast Track 100*
*Ironically, the Company's rapid expansion during the three years to 2002, which nearly brought the Company down, allowed Loganair to achieve position 84 on the Annual League Table of Britain's* ***Fastest Growing Unquoted Companies*** *in the 2002 Virgin Atlantic Fast Track 100.*
*Achieving position 86 and 77 on the Annual League Table of Britain's Private Companies with the* ***Fastest Growing Profits*** *in the 2008 and 2009 Sunday Times Profit Track 100 was much more pleasing.*

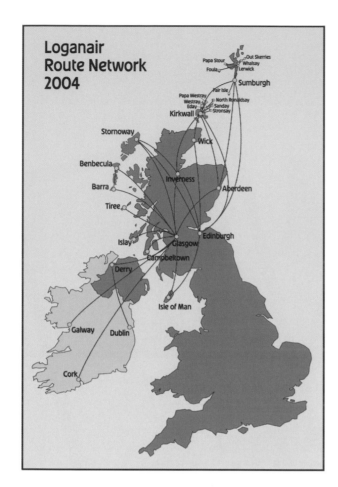

**Loganair Route Network 2004**

*2004 - The year Loganair made a breakthrough. At last British Airways had withdrawn from its remaining Highlands and Islands routes. The higher density Glasgow–Stornoway, Glasgow–Benbecula, Aberdeen–Sumburgh and Aberdeen–Kirkwall services were taken over by Loganair. By operating the smaller thirty-four-seat Saab 340B aircraft instead of the sixty-six-seat BAe ATP aircraft formerly used by British Airways, Loganair was able to increase the frequency and achieve double digit growth on these routes.*

*For the year ended March 2004, a profit of £765,000 was achieved on a turnover of £30.3 million with passenger carryings of 327,588. The aircraft fleet comprised seven Saab 340Bs, one Saab 340A, two de Havilland Twin Otters and five Britten-Norman Islanders. During the transition, three BAe ATPs were wet-leased from British Airways before being replaced by Saab 340B aircraft.*

BRAL were amalgamated with British Airways Regional and it had decided to give up the Scottish routes with a view to reducing its considerable losses. To allow the route transfers to take place, in March 2004 the Company leased four of British Airways' BAe ATP aircraft and crews until May 2005, supplemented by two wet-leased Fokker 50 aircraft and crews from Denim Air for ten weeks. By May 2005, Loganair had the necessary three Saab 340Bs in place with crews now trained, to give it a fleet of eleven Saab 340Bs and one Saab 340A aircraft.

Loganair was now operating the thirty-four seat Saab 340Bs on routes which had been served by BAe ATPs. Although previously highly unpopular with many passengers because of its earlier technical reliability record, the ATP did have the perceived benefit of sixty-four seating capacity. Initially, there was much criticism of the Saab and its very limited overhead luggage bins in the cabin. Never had the BAe ATP received greater acclaim; there was no mention now of 'Another Technical

Problem'. However, Loganair, by using the smaller and less-expensive Saab 340 was able to introduce increased frequency of service on the new routes and was rewarded with significant passenger growth. In the last years of British Airways operation, the Aberdeen-Sumburgh and Glasgow–Stornoway routes had been in serious decline. In the first year of its service, Loganair achieved double-digit passenger growth. The Company's increased-frequency strategy with smaller Saab 340 aircraft was beating the big aircraft capacity lobby hands down.

### Air Ambulance

It had become apparent long before the Tender for the Scottish Ambulance Service contract, commencing 1 April 2006, that Britten-Norman aircraft would not be acceptable for future air ambulance purposes. The Islander, without doubt, was the cheapest ambulance aircraft available and, with its outstanding short-field performance, could operate into most airfields. However, it was not pressurised, was very slow and had a cramped cabin. Of particular concern with the Islander were the patient-loading difficulties, which in today's Health and Safety culture were now unacceptable.

The new tender stipulated two fixed-wing aircraft, instead of three as formerly, and these were to be pressurised and based in Glasgow and Aberdeen. Two helicopters, as previously, would be required and based at Glasgow and Inverness. It was immediately apparent to Loganair that with four, and not the previous five, units being deployed, Orkney, which since 1967 had had access to a locally-based ambulance Islander, would not be as well served in future and greater response times were likely.

Loganair opted to tender with two Jetstream 31 aircraft, converted to appropriate ambulance interior configuration. The J31 was quick, pressurised and with a spacious cabin. Loganair's plan was to have a fleet of five J31s and it would use two for scheduled service purposes, have two dedicated to the ambulance role, and retain one as a maintenance back-up aircraft. Economies of scale would thus be achievable, which would give the Company a competitive advantage and allow it to submit a keenly-priced tender. As an optional extra for the Scottish Ambulance Service, the Company offered the part-use of one of its Kirkwall-based Islanders – to address the Orkney concern. This was quickly ruled out.

Competitive bid or not, the Company's tender was unsuccessful. Loganair management, pilots and staff were distraught. After all, many Loganair people had been so involved with, and had contributed so much to, the service since 1967. For them, there had not only been total commitment over the years, but a sense of carrying on a great tradition and of being part of the oldest air ambulance service in the world.

The key to successful flying operations has always been the suitability of the aircraft type for the task in hand. Loganair started to play a supporting air ambulance role to BEA in 1967 when it flew a small Piper Aztec to the Isle of Oronsay. The real impact and potential of Loganair's involvement was only realised, however, when the BN2 Islander was introduced and island communities, which had not been accessible to BEA's Herons, were now only a short flying time from mainland hospitals. Loganair took over the contract from BEA from 1973. The number of patients rose every year, and by the late 1980s, around 950 were being carried in a typical year.

By 1989, the Scottish Home and Health Department recognised that a major review of the Air Ambulance Service was long overdue. The Company was still notionally contracted to provide one Britten-Norman Islander in Glasgow on a twenty-four hour basis, 365 days a year. In practice, Loganair was making an Islander available in Lerwick and one in Kirkwall from 1700 hours to 0900 every night and all day Sunday, and would deploy whichever aircraft was nearest to the patient. Both aircraft in the Northern Isles were used in cases of emergency even when it meant cancelling or delaying scheduled passenger services. There was now a strong wish to put this on to a more formal basis, with greater control of the use of the service and, at the same time, the future employment of paramedics instead of volunteer nurses. More concerning from Loganair's perspective, helicopter trials were begun in Dundee. The status quo was being well and truly challenged.

As a direct result of the Review, when the new contract was awarded in 1993, the air ambulance service saw a major change with Loganair's fixed-wing aircraft being joined by the helicopters of Bond Helicopters Ltd. Dedicated Bölkow BO 105D helicopters operated from bases at Inverness and Prestwick while an Aerospatiale Dauphin 365 based at Plockton was available on a non-dedicated basis. At Aberdeen, Bond also provided a fixed-wing Beechcraft King Air for longer air ambulance flights.

From April 1993, as a result of the tendering process, Loganair continued to operate Islanders from Kirkwall and Lerwick. The Islander, for so long based at Glasgow, however, was not included. This was a great loss to the Company and a huge disappointment to the Glasgow-based pilots. Happily, normal service would be resumed seven months later and the Islander at Glasgow was restored after there had been complaints and concern expressed by doctors, and indeed the public, in the Hebrides and on the West Coast – all to the great relief and delight of the Glasgow-based pilots, for whom the ambulance role had become a way of life for so long.

There would be a permanent change for the nurses, however, who had also been vital to the success of the services from the 1930s. Except in Shetland, nurses were being replaced by Scottish Ambulance Service paramedics. Between 1938 and 1941, Peggy Boyd and Jean Govan of their own Paisley Trained Nurses Association were the nurses for air ambulance flights from Renfrew. From November 1941, the

*The Company's Britten-Norman Islander aircraft undertook air ambulance services from 1967 until 2006. Islander aircraft were effectively on duty round the clock at Loganair's Glasgow, Kirkwall and Lerwick bases. In 2006, the fixed wing responsibility of the Scottish Ambulance Service contract was taken over by the Super King Air aircraft of Gama Aviation.*

Photo: Tony Cowan

Glasgow air ambulance service was supported by nurses from Glasgow's Southern General Hospital. In Orkney, nurses came from Balfour Hospital and in Shetland from the Gilbert Bain Hospital. All the nurses received special training for their air ambulance duties which were performed on a voluntary basis, often during their off duty time.

After completing ten air ambulance flights, each nurse was awarded a Silver Wings brooch which was the only decoration not awarded by the nursing college which was permitted to be worn in the hospital ward. It was always worn with pride. Over the years, many nurses completed several hundred flights.

In 1976, a special award of a pair of golden wings was made to Nurse Gisela Thürauf from Glasgow's Southern General Hospital to commemorate her five hundred flights. The presentation was made by Captain David Barclay, himself a BEA veteran air ambulance pilot. Before she retired from nursing, Nurse Thürauf had completed more than nine hundred ambulance flights and was awarded the Queen's Commendation for 'Valuable Services in the Air'. Nurse Thürauf represented a breed of brave and selfless nurses without whom the service would not have developed. In 1991, they were replaced by highly trained specialist paramedics who were now accompanying patients in aircraft with the latest life-saving equipment.

From 1993, the Loganair pilots, at the three ambulance bases, were thus able to continue their ambulance flying duties through until 2006, very happily carrying on a tradition established by the early pioneers, flying often in challenging conditions when human lives are at stake. Countless families throughout the Scottish Highlands and

*Air Ambulance Babies*
*Baby Lisa Kerr, Loganair's*
*eighteenth inflight birth occurred*
*over Dunoon on the way from Islay*
*to Glasgow. Proud mother*
*Margaret Kerr was attended by*
*Nurse Jennifer McPhail.*

Courtesy of Loganair

*Captain Eddie Watt*
*looks almost as proud*
*as twins David and*
*Lynsey Henderson born*
*forty miles apart.*

Courtesy of
Captain Eddie Watt

*The first of Loganair's twenty-two babies born on board the air ambulance Islander was Katy Ferguson Leynair Devin on 2 August 1973, 2,000 feet above Kirkwall. Her name Leynair was a combination of the last syllable of the Captain's name, Captain Jamie Bayley, and the last syllable of Loganair. When Jonathan Ayres was born en route from Islay to Glasgow on 6 January 1982, his parents could not resist naming him Jonathan Philip Logan Ayres. For good measure, there were also twins David and Lynsey Henderson born on the way from Lerwick to Aberdeen on Friday 13 August 1982 – some forty miles apart! All 'airborne' babies were a source of great joy to everyone in the Company and Loganair presented them with an inscribed silver goblet to commemorate the event. Members of a very elite club.*

Islands had good reason to be grateful to the Loganair pilots and those many mercy flights. There was not one of the many Loganair pilots involved in the air ambulance service for the thirty-eight years who was not prepared to act above and beyond the call of duty. Many undertook a great many missions, but there were many who were in a position to make exceptional contributions. In the early days of Loganair's involvement, they included Captains Duncan McIntosh, Ken Foster, Bill Henley and later David Marris and David Dyer. In Orkney, Jim Lee and Andy Alsop, and later Stuart Linklater; in Shetland Alan Whitfield, Malcolm Bray, David Edmondston and Eddie Watt. And so many more.

Sadly, two Loganair pilots gave their lives while undertaking ambulance duties. In May 1996, the Shetland ambulance Islander, returning from Inverness to Lerwick, crashed about one mile south of Tingwall Airport. Captain Alan Young was killed. Miraculously, Doctor Gerald Freshwater and Nurse Maureen Polson were able to escape. In March 2005, Captain Guy Henderson, and paramedic John McCreanor, flying from Glasgow to Machrihanish to collect a patient, were lost west of the Mull of Kintyre on the approach to the airfield.

Photo: Steve Welsh, courtesy of Loganair and Beattie Communications

*31 March 2006, the last day of the Scottish Ambulance Contract was a sad day for Loganair's Glasgow-based staff who, apart from an interval of a few months in 1993, had been involved in air ambulance work since 1967. There was a real bond with the paramedic team which was also coming to an end. The Loganair staff, left to right, Captains Paul Deakin, Alex Brand, Bryon Smee, Graeme Abernethy, Andy McKague and George Cormack, and Loganair's Engineering Supervisor, Michael Kaloheris. On the left is Jim Cameron, Loganair's Managing Director from 2001 to 2006. The author is on the right.*

By 2006, around a thousand patients were being carried annually on Loganair's three ambulance Islanders. All patients are important, but none more than the twenty-two babies, including the Henderson twins from Shetland who were born on board the Islander some forty miles apart. The whole Company shared the joy of having air ambulance babies. It was the Company's practice to present each baby with an inscribed silver goblet to commemorate the event. A most exclusive club.

For almost forty years, ambulance flying for the many Loganair pilots could be routine, or challenging, or downright dangerous, but it was always a source of immense satisfaction. The air ambulance baton was handed over to Gama Aviation, a Super King Air operator based at Farnborough, and Bond Aviation Services who had been contracted by Scottish Ambulance Service since 1991.

### Shetland

For the Company management and staff, the loss of the Scottish Ambulance Service contract in 2006 was certainly emotional, but it also had serious financial implications for operations at the former Islander ambulance bases. Glasgow, with its many aircraft and scheduled services, of course, was least affected, but in Kirkwall the withdrawal of one of the two locally based Islander aircraft was more serious. It was the Shetland base that was worst affected. In Lerwick the same single Islander aircraft undertook both ambulance and local scheduled service flying and the loss of the ambulance contract impacted hugely on the economics and the viability of the inter-island air service.

On the passenger service front, the closure of RAF Saxa Vord in Unst, just like the rundown of RAF Machrihanish in the Mull of Kintyre, and Benbecula in the Western Isles, the so-called 'peace dividend', had affected the local air services in recent years. For many years the military base not only contributed significantly to the economy of the island, but had generated much passenger demand for the local air services. Indeed, for several years Loganair had a contract with the MOD to carry military personnel from Edinburgh to Unst via Lerwick. Shetland's inter-island Tingwall–Unst and Sumburgh–Unst services benefited significantly from military personnel usage. By 2006, there was no Unst service and the SIC Contract Tender related to local passenger air services linking Tingwall with Foula, Fair Isle, Out Skerries and Papa Stour, and between Sumburgh and Fair Isle. Total passenger carryings, now without Unst, were much lower than in the heydays of the 1980s and 1990s.

By 2006, the Scottish Ambulance Service contract was bearing sixty percent of the cost of aircraft, pilots and local overheads of Loganair's Lerwick base primarily because of the twenty-four hour ambulance cover being provided. Clearly, when the ambulance role ended, all Lerwick costs in future would have to be charged to the Shetland Islands Council's Public Service Obligation contract for the local internal air

> Fair Isle Primary
> Fair Isle
> Shetland
> ZE2 9JU
> 01/07/05
>
> Dear Eddy,
> Thankyou for flying us to Shetland and back. It was a shame we got stuck on Wednesday but we were glad when we got back on Thursday. Thanks for getting us back on Thursday. We thought we were going to get stuck till Friday.
> yours Sincerely
> Erin Welch

*Letter sent from Fair Isle Primary School to Captain Eddie Watt, Loganair's Senior Pilot in Lerwick, Shetland – clearly during a period of typical Shetland summer weather. Could we not have more customers like this young lady – understanding, reasonable and wholly appreciative of Loganair's service.*

services. It was obvious that the next Tender Price for the PSO contract, also scheduled for 2006, was going to be significantly higher.

With a view to achieving economies of scale, SIC held discussions with Orkney Islands Council whose own PSO Contract Price would also inevitably increase as a consequence of Loganair no longer having an ambulance Islander based at Kirkwall Airport. The two Islands Councils clearly had common purpose. Orkney thought that joint operations would result from the discussions, but SIC eschewed any possibility of economies of scale and, to the great surprise of Orkney Islands Council, signalled an independent course of action by acquiring two Islander aircraft for its own Shetland services.

After the early difficulties of the 1970s, SIC, and the Chairman of the Transportation Committee, Jim Irvine, had been most supportive of Loganair on many occasions. Awarding the Company the Oil Pollution Surveillance Contract at Scatsta near Sullom Voe Oil Terminal had helped stabilise the flying operation in Shetland. Along with Orkney and the Western Isles, the Council had sponsored a local pilot during a national pilot shortage. When Loganair had applied for a licence to operate on the Aberdeen–Lerwick route, Councillor Jim Irvine agreed to appear as a Loganair witness at the Hearing. By the time of the 2006 PSO Contract Tender, however, the Loganair management were surprised and saddened to form the clear impression that SIC were looking for change, for a new Tingwall-based operator. Nonetheless, Loganair, despite having no ambulance role in Shetland, submitted a keenly-priced Tender, but the Company, and indeed the Tingwall staff, learned from the *Shetland Times* and *Shetland News* that it had not been accepted, apparently because of price. The Council repeated the process, but the second Tender failed procedurally.

By the time the Council had prepared and advertised its Tender for the third time, the Company had decided not to submit another Tender. For the first time since Captain Alan Whitfield had first landed in Shetland in 1969, Loganair would not have a locally-based Islander aircraft.

It was not only Loganair's Tingwall based staff who were bitterly disappointed at this turn of events. There were many Loganair 'soothmoothers' at the Glasgow Headquarters who had played their part over the years to ensure that the Company's furthest flung base had the necessary engineering or pilot or even management support. Thirty-seven years in Shetland surely merited an appropriate farewell, and the author took it upon himself to invite Alan Whitfield, who had undertaken Loganair's first Shetland flight, and Captain Eddie Watt who would fly the last Shetland service, to accompany him and Gordon Young, the Company's Director of Flight Operations, on a tour of the islands to say 'thank you' and bid a proper 'farewell' to the local communities.

On a truly beautiful, sunny day the Loganair Islander flew to Papa Stour, then on

*What security? The wonderful buffet lunch party held around the aircraft on the airstrip at Out Skerries and given by the islanders when Eddie Watt, the last Loganair base captain, and Alan Whitfield, the first Loganair base captain, accompanied by Gordon Young, Loganair's Director of Flight Operations, and the author, made their farewell tour of the Shetland Islands in July 2006.*

Photo: Alan Whitfield

to Out Skerries where Alice Arthur and almost every person on the island provided a splendid buffet lunch right there on the airstrip beside the aircraft. The party then flew to Foula where Isobel Holbourn had arranged for the school children to display Loganair photographs and memorabilia, and then finally on to Fair Isle and the biggest reception of all organised by Fiona Mitchell and Dave Wheeler.

It was a day of sheer nostalgia tinged with more than a little sadness by reminiscences and great memories of the many Base Captains who had become household names: Ian Ray, Malcolm Bray, Ian Potten and the Laird of Unst, David Edmondston. There was no doubting the goodwill and genuine affection with which all the Loganair pilots were held by all the Shetland island communities, but it was the end of an era. A little comfort was taken from Shetlander Eddie Watt's comment that, 'Hit might be da end o an era, but da peerie plane ill aye be minded as da Loganair.'

### Fortieth Birthday Party in Orkney

The Orkney Inter-Island service is viewed with great pride and affection as it was the Company's first bit of business of real substance, and anniversaries are marked and celebrated. On the Tenth Anniversary in 1977, Loganair's Birthday Cake was cut by the Conservative Leader of Her Majesty's Opposition, Mrs Margaret Thatcher, who happened to be campaigning in Orkney. She certainly entered into the spirit of the occasion and personally offered slices of the cake to all the oil workers who were waiting for their helicopter transfers in the Kirkwall Terminal Building, giving as good as she got in the verbal exchanges with them. Also present that day was Desmond Norman who, along with John Britten, had designed and built the Islander aircraft at Bembridge, Isle of Wight. The great enthusiasm in Orkney for the Islander and its obvious suitability for the local North Isles operation must have gladdened his heart.

Ten years later, the Anniversary Dinner was held in the Merkister Hotel in Orkney with Lord Grimond, former MP for Orkney and Shetland, as principal guest of honour. It was he who had once famously said in a transport debate in the House of Commons that the nearest railway station to his Shetland constituency was in Bergen. At the Twentieth Anniversary Dinner in the Merkister, Lord Grimond famously had said nothing. Indeed he had said nothing at all throughout the meal despite the valiant efforts of the author's wife to engage him in conversation. Nothing at all, that is, until it was announced that Michael Bishop was being invited to speak on behalf of the Company. 'Oh, I want to hear him', declared Lord Grimond, and took his hearing aid out of his jacket pocket. It is not an exaggeration to report that a most distinguished political career nearly ended at that precise moment.

The Thirtieth Anniversary was very special because it took place shortly after the

*In September 1977, the 'Iron Lady', Mrs Margaret Thatcher, then Leader of Her Majesty's Opposition, was campaigning in Orkney on the day the Company celebrated the tenth anniversary of its Orkney inter-island air services. She readily cut the birthday cake and handed it round the many oil workers who were in the terminal building.*

Courtesy of Orkney Library & Archive

'new' Loganair was up and running and the Dinner somehow also marked a new beginning. Among the guests in the Kirkwall Hotel was Alan Bullen who had been so directly involved at the very beginning of the North Isles services as General Manager of the Orkney Islands Shipping Company under whose auspices the services operated for nearly the first ten years.

The Fortieth Anniversary had to be even more special. On 27 September 2007, the Company arranged a Dinner in the St Magnus Centre in Kirkwall. It was an evening of celebration, made all the more remarkable by the number of guests who not only had been directly involved in the service, but had actually been active throughout the forty years.

It had been the arrival of the Britten-Norman Islander that had allowed Loganair to re-establish the air services which Captain Fresson had operated in the 1930s until ended by the outbreak of the Second World War. In 1967, two of Fresson's pre-war airstrips, on North Ronaldsay and Sanday, could still be used, but new airstrips had to be established on the other islands. There was a last minute hitch when the Scottish Civil Aviation Department decided all the island airstrips required to be lengthened. Captains Ken Foster and Jim Lee carried out route certification while the old Orkney County Council worked frenetically to extend landing distances from 1,500 feet to 1,800 feet in an impressive four weeks. These grass airstrips provided the air service infrastructure for many years until the Council upgraded them to hard runways in 2003/4.

At the island airports were many attendants whose commitment to the air service contributed greatly to its success, and many were among Loganair's dinner guests.

They included Mrs Nan Scott, widow of Councillor Jack Scott who wrote *Wings over Westray* and who had attended Loganair's Westray flights from the very first day for nineteen years before handing responsibility over to her daughter Mrs Linda Hagan and her husband Stephen to continue the family tradition. Also present were Jim Lennie who had been airfield attendant on Sanday throughout the forty years, and Billy Muir at the North Ronaldsay airstrip and Bobby Rendall in Papa Westray – all for many years the unsung heroes of the inter-island air service.

Many Orkney pilots were present. Andy Alsop and Jamie Bayley from the early years, Mike Seyd and David Kirkland from more recently, but with no less than fifty years inter-island service between them. Captain David Kirkland would later be awarded an MBE in the Queen's Honours List for Services to Aviation in Orkney. The initial service from 1967 had linked Kirkwall with the North Isles of Stronsay, Sanday, North Ronaldsay, Westray and Papa Westray, and later Eday. Captain Andy Alsop in particular had been a driving force in the development of the inter-island air services. The South Isles of Flotta and Hoy were added, but both services were discontinued in 1981, the latter unsustainable in the face of competition across Scapa Flow from ferries provided free by Occidental Oil.

Courtesy of Loganair

*Captain Stuart Linklater and Phyllis Towrie welcome regular traveller Jim Meason. Orkney's North Isles services provide an essential transport system used regularly by bankers, vets and school teachers. Until she retired in 1986, Maisie Muir, who worked for the Royal Bank, had undertaken an estimated 8,400 flights, but that astonishing number could still be overtaken by her successor, Anne Rendall. Jim Meason, school hostel warden, by his own estimation, since 1977 has clocked up enough miles around the North Isles to equate to travelling four times around the world.*

As a result of the efforts of a great many people, passenger carryings increased every year reaching a peak in 1989/90 when some 26,000 passengers travelled in the single, eight-seat Islander, an extraordinary performance and flying activity. This was in the year immediately preceding the introduction of the roll-on roll-off ferry services in the North Isles. Not surprisingly, annual passenger traffic immediately plummeted to 12,000. North Ronaldsay and Papa Westray were left out of the roll-on roll-off ferry programme and air passenger traffic levels from these two islands continued unabated. Again by considerable effort, passenger carryings would increase steadily for the next twenty years to an annual traffic of more than 20,000 – just about the same level as the entire population of the Orkney Islands.

The Orkney Inter-Islands Service has been a Loganair team effort. Not only pilots, but a host of engineers and customer services staff. Mrs Christine Allan, with no less than thirty-two years service, was present at the dinner as was Bryan Sutherland who has been maintaining Loganair aircraft for all of forty years and based at Kirkwall Airport except for a few months which he served at Loganair's base in Stornoway.

Loganair's North Isles services are an essential part of the transportation system. Itinerant teachers, vets, the school dentist, the chiropodist, the telephone engineer, and of course, the banker, are regular passengers. Miss Maisie Muir who worked with the Royal Bank had an estimated 8,400 flights with Loganair between 1970 and 1986, but that astonishing number could be overtaken by her successor, Miss Anne Rendall, who

Courtesy of Loganair

*By 2006, the Britten-Norman Islander was being deployed only on the inter-island services in Orkney. A commercial arrangement was agreed with Highland Park, distillers in Kirkwall, which entailed both Orkney-based Islanders being repainted in the Highland Park livery.*

was at the Dinner, or by fellow guest Jim Meason, who by his own interesting calculation had flown the equivalent of four and a half times round the globe in his flights with Loganair around Orkney as warden of the school hostels.

In view of the Company's close collaboration, it was hardly surprising that the Orkney Islands Council was well represented at the dinner by many who had been involved with and had supported the inter-island air service over the years. Like every partnership, there had been many difficult issues to be resolved during the forty years, but the relationship was remarkable for never having had any personal rancour. In the early years, Convener Edwin Eunson and Chief Executive Graham Lapsley had been supporters, but present at the dinner were many others who had played a positive role, including ex-Convener Hugh Halcro-Johnson, George Stevenson, Jim Sinclair, Bob Sclater and Convener Stephen Hagan. Everyone in their different ways had helped to make the inter-islands air service, not only a great success, but also a model of its type.

Loganair's base in Kirkwall received visitors from many countries coming to see for themselves how the operation worked. The author was once asked to speak live on *Radio Tokyo* for the benefit of the 'many listeners who were apparently especially interested in the Westray–Papay two minute route.' 'No. Champagne and caviar were not on the menu that day. And, no, the in-flight movie was not *Gone With The Wind*.'

### The vexed question of air fares

'It is cheaper to fly to New York than it is to fly to Shetland' is a caustic comment that has often been made against the airlines. There has been a variant on this with the increasing cost of fuel: 'It is cheaper to fly to Sydney than it is to fly to Shetland.' Since the 1960s, it seems that nothing was likely to raise emotions more than the subject of air fares. Services to the islands are vitally important to the social and economic welfare of the remoter communities, and the cost of travel whether by air or sea is necessarily a most sensitive issue. Councils and consumer groups have regularly protested or complained, whether it was BEA, or later BAHD, or indeed Loganair that was involved. Earlier in the tariff-regulated regime that existed until 1986, formal objections to airline fare levels were routinely lodged. Such was the strength of feeling, island council representatives would often travel all the way south to attend CAA Tariff Hearings in London.

In November 2001, the author presented a paper at an important Department of Transport Conference in Edinburgh on an *Air Transport Strategy for Scotland*. In his opinion, most of the main complaints by air transport users in the Highlands and Islands for the past several decades had been addressed. Service reliability was much improved by the introduction of more modern aircraft using airports equipped now with better navigational aids. The use of slow, unpressurised aircraft was now a thing of the past; service frequency had been increased across the network, made possible by the replacement of unnecessarily big, sixty-four seat aircraft with smaller thirty-four seat Saab 340 aircraft. All well and good. Loganair, however, could not claim to have resolved the most important issue – the high cost of air fares on the Highlands and Islands routes. He readily acknowledged that being able to offer affordable air fares was the key to a successful air transport system in Scotland, but expressed the view that air fares could only be reduced in any significant way by Government intervention, perhaps by providing additional subsidy to HIAL to allow it to reduce or eliminate its now very onerous charges to passengers and airlines on the Highlands and Islands routes. He also argued that Public Service Obligation (PSO) status may be extended to a few more routes as a way of reducing fares. In Scotland, historically, whether by the Scottish Government or the Islands Councils, the PSO mechanism had only been used in cases where the route would not be sustainable without financial support, but never as a mechanism to reduce fares.

Without subsidy by whatever mechanism, air fares in Scotland were destined always to be high. Costs were high and passenger volumes and revenue low. Aircraft were expensive and doing relatively little flying, rather less than half the daily flying hours of the low cost carriers with whose fares the Scottish airlines are invariably compared. Flights are short, causing airport charges to impact disproportionately on airline costs. Fuel costs are volatile and high, and much higher again if the airline has to uplift kerosene at Sumburgh, Kirkwall or Stornoway airports where prices can be nearly three times as high as at Glasgow or Edinburgh. To combat this and to keep fares as low as possible, Loganair had been extending the range of discounted fares and had managed to stimulate passenger growth, but this was not enough.

In 2003, a number of Regional Transport Partnerships were set up by Scottish Transport Minister, Tavish Scott, MSP for Shetland. HITRANS had the responsibility for Argyll and the Islands, Comhairle nan Eilean Siar, Highland Region and Orkney. Shetland had a separate Authority, ZetTrans. HITRANS engaged consultants who produced a report and a scheduled service operating model for the Highlands and Islands with all routes the subject of PSO contracts and designed, no doubt, to satisfy the varied interests of the many members of HITRANS and ZetTrans. The entire scheme, estimated to cost the Government £12 million, would be based on fares reduced by thirty percent from their current level.

In Loganair's opinion, the consultants proposed scheduled service network was seriously flawed conceptionally and operationally, with costs grossly understated and passenger revenue generation excessively optimistic. Loganair did speak from a position of some strength as it had been operating many of these services for many years and knew the actual costs involved. Nevertheless, for some time the recommendations of the HITRANS Report were seized by many parties as a panacea. There was growing local Government support for PSO status being applied to more, if not every, route in the Highlands and Islands and the ruling Labour/Liberal Democrat Coalition at the Scottish Parliament in their Partnership Agreement committed the Scottish Executive to consider the proposition during the life of the 2003/07 Parliament.

Later the following year, 2004, the HITRANS recommendations were indeed reviewed in the context of the Comprehensive Spending Review. The Scottish Executive seemed to share Loganair's serious misgivings about the cost of the whole project and raised considerable doubts about whether sufficient funds were available to fully fund the project. In any event, new European legislation was making it highly unlikely that multiple Public Service Obligation contracts would be approved. Perhaps because of abuse elsewhere in Europe, PSOs in future would be granted only 'sparingly' and only in instances where a lifeline route would not survive without the financial support of a PSO mechanism. The HITRANS proposition, with just about every air service in the Highlands and Islands affected, would clearly not qualify.

Instead, the Minister of Transport at the Scottish Parliament, Tavish Scott, introduced the Air Discount Scheme (ADS) in May 2006 under the *Aid of a Social Character* mechanism. The European Commission approved the scheme as a means of improving connectivity and social inclusion for those registered residents in designated areas of Scotland. Members benefit from a forty percent discount on air fares, which the airlines recover from the Scottish Government monthly in arrears. Designated areas are Orkney, Shetland, the Western Isles, Islay, Caithness and Sutherland.

The Air Discount Scheme has been a great success. Such is the price sensitivity in the Highlands and Islands that, between 2005 and 2011, passenger carryings on the ADS-designated routes actually rose by twenty percent. By contrast, during the same period, passenger volume on UK domestic routes fell by twenty percent. This has been mirrored on routes within Scotland on which ADS does not apply where passenger carryings also fell by twenty percent. Since April 2011, however, when the Scottish Government introduced a restriction on business use, the number of passengers using the Scheme has dropped by up to a third on some routes.

### Diversification

Loganair had been given more than a year's notice that the Company's contract with the Scottish Ambulance Service was not being renewed in 2006 and the Board realised that the Company's business would then be limited to scheduled passenger services, which in the main were in the Highlands and Islands. 'Eggs in one basket' came to mind. Indeed, by this time, Loganair operated the large majority of the air services in the Highlands and Islands and recently had been working hard to increase the frequency of service on many of these routes. The Board realised there was not much more scope to expand further in its Highlands and Islands heartland and took the important strategic decision that the Company had to diversify. This meant trying to win new air freight and air charter work and, in the case of scheduled services, identifying routes outwith the Highlands and Islands.

The Company's core business was, and would continue to be, the provision of scheduled services in the Scottish Highlands and Islands. Some of the longest established air services, such as from Glasgow to Aberdeen, once one of British Airways' busiest routes, and from Glasgow to Inverness, were being discontinued at this time. Passengers were simply finding that travelling by road or rail for these distances was now much cheaper and easier. Identifying new scheduled service routes outwith the Highlands and Islands, however, was a challenge, and somehow reminiscent of Loganair's great pioneering days of the 1960s and 1970s. Route evaluations by management in 2005 and 2006 were perhaps a little more sophisticated than those of thirty or forty years earlier, but were still inevitably to a degree 'trial and error'.

In Ireland, the Company commenced services from Glasgow to Cork and to Galway, thus adding to Loganair's existing presence on the Derry–Dublin PSO route. Later, new services were launched from Aberdeen to Cardiff. On the Isle of Man, services were started on the Belfast and Blackpool routes. Sadly, due to low passenger demand or an inability to achieve an adequate fare, or a combination of the two, all of these new services had to be withdrawn after a relatively short period. Plus ça change.

The Company's commitment to Dundee, however, saw new services to Birmingham and Belfast launched and making steady, if unspectacular, progress. In summer 2008, services from Sumburgh to Bergen were started successfully following the withdrawal of the Smyril Line ferry service between the two points. A Kirkwall–Bergen summer service was added the following year.

*Water cannon welcome at Bergen for the inaugural Saab 340 flight of Loganair's new scheduled service from Sumburgh to Bergen in 2007.*

Courtesy of Loganair

Increased efforts were also being made to capture more of the ad hoc charter market, and, in particular, the growing fixed-wing support for the North Sea Oil and Gas business. Aberdeen once again began to assume a greater importance for Loganair. When an Icelandic airline, City Star Airlines, went into receivership early in 2008, Loganair acquired its wholly owned subsidiary, Caledonian Airborne Engineering Ltd (CAEL), believing that the hangar and engineering facilities at Dyce Airport, Aberdeen, would strengthen the Company strategically. CAEL would not only provide local support for Loganair's two most important scheduled services from Aberdeen to Kirkwall and Sumburgh, there would also be a new income stream from third-party engineering.

The engineering company acquisition was not an immediate success and required a name change to Aero Handling Ltd and a reorganisation with the emphasis now on ground handling services at Aberdeen. In 2011, the ground handling services at Sumburgh Airport were taken over from GAC Shipping Ltd (GAC) and were now operated by Aero Handling Ltd. Shetland continues to be extremely important to Loganair because of the Company's scheduled services to Aberdeen, Edinburgh, Glasgow, Kirkwall and Inverness, but important too because of the Company's increased involvement in oil and gas industry support.

If the Loganair management was turning the clock back in chasing North Sea Oil work, there was a real sense of déjà vu when three new mail contracts were secured. Two Saab 340A aircraft were acquired for the two night contracts for Royal Mail from the Aberdeen base, while a third aircraft was leased in from Lithuania for a daytime mail contract from Inverness. Later, once HIAL was persuaded in 2010 to open Dalcross Airport at night, Loganair was also able to undertake all three Skynet Contracts with its own two Saab 340 aircraft. When Highland Airways went into administration in early 2010, the Company initiated a newspaper contract operating between Aberdeen and Stornoway, allowing the two Saab 340s to be fully committed by night and day.

As well as the night contract charter work for Royal Mail, Loganair routinely carries mail on many of its scheduled flights. Deliveries to both Islay and Tiree being entirely dependent on the morning flights from Glasgow. A slightly more unusual event occurred in August 2010, when Janet MacLean, the Company's Station Manager in Barra, called to say that a man had appeared at the airport shortly before the departure of the flight to Glasgow with 'the Royal Mail'. As the telephone call progressed, it became obvious that the man was a Queen's Messenger and the Royal Mail in question was nothing other than Her Majesty's private dispatches, sent whilst she was holidaying in the area aboard the *Hebridean Princess*. Captain Fraser Beaton was quickly instructed to guard with his life the sealed mail bag. As the Twin Otter lifted off from the beach at Barra, a volley of urgent calls were made to the Royal Mail: 'What's to be done with the bag when we get to Glasgow?' The quandary was soon resolved. As the Twin Otter taxied in an hour later, the Postmaster General for the Glasgow district was standing to attention by the allocated parking stand to collect the mail bag – attired in full ceremonial uniform! Loganair once again had performed its duty to Queen and Country.

Still pursuing its policy of diversification, and reducing its dependence on Highlands and Islands scheduled services, in July 2011 the Company acquired Cambridge-based Suckling Airways, which formerly traded as Scot Airways. Suckling by then was specialising as an ACMI, wet lease provider for a range of airlines. It had also built up a loyal customer base for its charter operations, notably for many English Premiership Football Clubs. Loganair was attracted by Suckling's fleet of five Dornier 328 aircraft whose superior speed and range would allow the now Loganair Group to have operational and commercial options not available with its Saab 340 fleet. Loganair now had another string to its bow. Suckling retained its own Air Operator's Certificate (AOC) and became a wholly owned subsidiary of Loganair.

*In July 2011, the Company acquired Cambridge-based Suckling Airways which had previously traded as Scot Airways. Suckling's fleet of five thirty-one seat Dornier 328 aircraft are used on charter operations and for services on behalf of other airlines.*

Photo: Iain Hutchison

The strategic decision to diversify into a range of activities and reduce its dependence on the Highlands and Islands scheduled services has been vindicated. Even during the economic recession in 2008, 2009 and 2010, the Company returned a profit which made it almost unique in the UK airline industry. In fact the Company, in its fiftieth year, achieved profitability for the eleventh consecutive year and a stronger balance sheet than at any time in its long history. By now, Loganair, the airline, has an annual turnover of around £65 million and has some four hundred staff. The Company operates a fleet of twenty aircraft comprising two Twin Otters and two Britten-Norman Islanders as well as the main fleet of Saab 340 aircraft of which ten of the sixteen are owned and all without the Company incurring any debt. It is a solid financial platform on which to face the endemic vagaries and uncertainties of the airline industry.

## Franchises

In 1993, British Airways granted a franchise to Gatwick-based, CityFlyer Express. At that time, protracted discussions were taking place between British Airways and Loganair about the feasibility of a merger or a joint venture. This did not happen, mostly because of the competition that existed between British Midland and British Airways, but a compromise was reached for Loganair to become British Airways' second franchise partner. Under the guidance and drive of Lewis Scard, British Airways' new Franchise Manager, many more franchises were awarded and Manx Airlines (Europe) quickly followed Loganair into the British Airways franchise fold and in a few years were joined by GB Airways, Maersk Air, British Mediterranean Airways, BASE Airlines of Eindhoven, Sun-Air of Scandinavia and Comair in South Africa.

Photo: Kieran Murray

*Even the smallest aircraft type in the fleet, the Islander, was permitted to be part of the British Airways franchise operations from 1994. The variety of 'ethnic' tails is seen here at Sumburgh Airport. This British Airways livery would be ended famously with Mrs Thatcher covering an aircraft model with her handkerchief to show her displeasure. At the Franchise renewal in 2003, the Islander services were excluded because they were single-pilot operations and the Islanders reverted to Loganair's own livery.*

British Airways' motivation and objective was to maximise interlining connections and feeder traffic for British Airways' mainline flights. For British Airways, franchising was a most effective strategy. Much was made of the so-called 'Heineken effect' with the franchise allowing British Airways to reach airports that it could not reach itself. For Loganair and the other franchisees, this was of immense benefit to its customers. In the Highlands and Islands passengers could now check themselves and their luggage through from, say, Kirkwall to Edinburgh on a Loganair flight before connecting with a British Airways flight to London Heathrow and onto any destination on British Airways' worldwide network. This feeder traffic was most worthwhile for both airlines.

All aircraft in the Loganair fleet were in British Airways livery. This included Loganair's Britten-Norman Islanders which operated the inter-island services in Orkney and Shetland. When, in 2003, the franchise renewal was being negotiated, British Airways determined that the franchise agreement would not extend to Islander operations because they were flown as single crew. The Islander aircraft in the Northern Isles were quickly painted in Loganair's own livery. All other Loganair aircraft were British Airways-branded.

By the time Loganair had renewed its franchise through to 2008, most of British Airways' franchise partners, CityFlyer Express, Manx Airlines (Europe), British Mediterranean, Maersk and BASE had fallen by the wayside. There was the clear impression too that relations between GB Airways and British Airways now were strained. British Airways' Chief Executive, Willie Walsh, expressed the view that franchises, certainly British Airways' UK franchises, had run their course, and Loganair's franchise agreement would not be renewed. Loganair believed that the Company had somehow been caught up in the crossfire between British Airways and GB Airways. Only Sun-Air of Billund in Denmark and Comair in Johannesburg in South Africa continued as British Airways franchise partners.

Loganair then had to consider the future. It could operate as an independent Company, in its own livery with its own brand. Although, for the previous fourteen years, it had been subsumed into British Airways' brand, Loganair still had a good, well-respected reputation in Scotland. Indeed, the author remembers vividly the dark, desperate days of the mid-1990s when he was asked at a meeting in Shetland if the Company would survive. He sincerely hoped so. 'Will the name *Loganair* survive?' Before he could answer, one of the audience shouted, 'Of course it will. My mother still shops at Liptons.'

Alternatively, the Company could enter a new franchise agreement with another airline and the author approached Flybe which was in the throes of taking over the many regional air services of British Airways Connect. Discussions and negotiations were protracted. The Flybe model was that of the low-cost carrier and very different from the 'legacy' carrier model of British Airways and Loganair. The Loganair Board took a long time to take the finely balanced decision to sign the Flybe Franchise Agreement to commence on 1 November 2008. Loganair's partner was now the largest regional airline in the UK and Europe and its passengers would have easier flight connections and greater choice of destinations.

*The British Airways franchise ended in 2008, to be replaced by a Franchise Agreement with Flybe, the UK's largest regional carrier. Here the Saab 340 flies overhead Dundee Airport, where the company developed the Riverside airstrip in 1963 during the construction of the Tay Road Bridge.*

Courtesy of Loganair

Flybe was not British Airways, however, and inevitably there were concerns raised in the Northern and Western Isles about this change to their air services. For the first time since the Second World War, there would be no British Airways 'presence' or identity in the Scottish Highlands and Islands. There would be no smiling cabin staff in the smart British Airways uniform to greet passengers on to the aircraft with the instantly and internationally recognisable British Airways tail. No matter that the success or failure of these routes had been totally the responsibility of Loganair, and not British Airways, since 2003. There was a sense of loss. After nearly forty years of worries and uncertainties about first BEA, then British Airways and then BAHD and eventually British Airways CitiExpress, British Airways had finally 'withdrawn' from Scotland.

Psychologically this was disappointing, but there would be also a more practical loss to the passenger, the connection with the British Airways worldwide network of scheduled services and destinations. The Company immediately set about negotiating and securing a Code Share Agreement with British Airways which crucially allowed Loganair's passengers to have all the services and connections they enjoyed hitherto under the British Airways franchise. This arrangement for the passenger represented the best of both worlds and Loganair was delighted. As well as being able to book on to Flybe's extensive network in the UK and Europe, the British Airways Code Share Agreement allowed the Loganair passengers anywhere on Loganair's network to check themselves and their luggage through to any destination of British Airways worldwide network.

Any change from British Airways to Flybe, or indeed any other partner Loganair may have chosen, would take some time to implement and settle down. Loganair management and staff worked patiently with passengers through the inevitably difficult transition period to explain the new systems and choices.

*The Flybe cabin staff uniform worn by Loganair's Dundee-based staff. The Company started scheduled services in 2008 to Belfast City and Birmingham from Dundee Airport. In 1963, Loganair had operated its first scheduled service from Dundee - to Turnhouse Airport, Edinburgh.*

Courtesy of Loganair

### Merry-Go-Round

For decades, it seemed, British Airways had been prepared to do whatever was necessary to resist the covetous attentions of other airlines, including Loganair, in order to retain control of its regional route network despite incurring huge financial losses in the process for most of that time. Suddenly, in 1996, British Airways handed over its remaining Highland Division services to British Regional Airlines (BRAL), which included, of course, the Scottish routes.

And then, in 2001, British Airways, through its subsidiary, The Plimsoll Line, bought Manx and BRAL and, in so doing, was buying back for £75 million the services which it had given away for nothing only four years earlier. The rationale, if there was one, was that newly arrived Chief Executive, Rod Eddington, believed that it was imperative for a big international airline like British Airways to have a 'presence' in the regions. In this, he may have been unduly influenced by the Managing Director of Brymon Airways,

Gareth Kirkwood, or perhaps even more so by the Brymon Chairman, Sir Frank Kennedy, who always wanted a British Airways presence at Manchester.

In any event, Manx and BRAL were duly absorbed into British Airways subsidiary, Brymon, and the new airline renamed British Airways CitiExpress. The timing was unfortunate in the extreme. Following the terrorist attacks of 9/11, British Airways CitiExpress required cutbacks and further reorganisation, including absorbing another British Airways subsidiary, British Airways Regional (BAR) and its Boeing 737 operations at Manchester and Birmingham. This four-into-one new airline, British Airways CitiExpress had a combined fleet of no less than eight-eight aircraft comprising eight different aircraft types. Many industry observers thought the plan hopelessly ambitious and they were proved be right.

Further cutbacks began almost immediately and the fleet was reduced to sixty aircraft. To achieve this, radical surgery was necessary. Loganair declined the invitation from British Airways to take over its fleet of eight BAe ATP aircraft and the routes which they operated. The Company, however, was only too pleased to take over British Airways' remaining internal Scottish services and entered a short-term wet lease arrangement for four of British Airways' ATPs while more appropriate aircraft, Saab 340Bs, were being sourced and introduced.

Lest we forget. Yes, these were the very same Scottish routes that British Airways had handed over to BRAL, which were expensively bought back by British Airways when acquiring BRAL four years later, and now were being handed over to Loganair – only thirty years or so after the Company had made its first overtures to BEA for them. At the same time, British Airways was handing over twelve Jetstream 41 and several routes to Eastern Airways, and their routes out of Plymouth and three Dash 8-300 to Air Southwest. British Airways' Isle of Man services were largely abandoned.

The reduced British Airways CitiExpress operation, however, continued to struggle, losing some £30 million per year. The Grand Strategy of rolling all the regional airlines into one business, and adopting BRAL pilot and cabin crew costs in Manchester, Birmingham and probably Glasgow to replace British Airways Mainline operations, was never going to be accepted by British Airways' unions. In 2006, there was further reorganisation and yet another new company, British Airways Connect was formed with an all-jet fleet of BAe 146 and Embraer 145 aircraft. There was to be no quick fix, however, and £6 million loss was incurred in the first six months. Very soon, negotiations were begun with Flybe, and British Airways Connect was handed over to Flybe along with a financial dowry of £140 million.

Out of its former comprehensive network of services, British Airways retained only the Manchester–New York JFK service which the old British Airways CitiExpress had been operating with a Boeing 767. Separately, a new subsidiary, British Airways

CityFlyer was set up to operate a range of services out of London City Airport with Avro Regional Jets, which were upgraded BAe 146 aircraft. Everything else had gone to Flybe in March 2007, and the rest before that to other airlines including Loganair.

It was not just the routes that had been on a merry-go-round. Some of the aircraft had too - for example, the Jetstream 41 aircraft. These aircraft were originally purchased by Loganair and Manx Airlines in 1994 as part of the Airlines of Britain sixteen-aircraft deal with British Aerospace designed to extricate Loganair from the disastrous BAe 146 contracts entered into ahead of the Company's application for the ex-BCAL Gatwick routes. British Airways became the, perhaps not so proud, owner of these J41 aircraft when it acquired Manx and BRAL in 2001. When reducing its operation in 2003, British Airways CitiExpress released these same J41s to Eastern Airways, requiring British Airways' Chairman, Rod Eddington, to note in his Chairman's Statement details of the transaction and the £27 million write-off. All a case of what goes round, comes around.

*The Highlands and Islands services were now firmly established with greater service frequency. The Company's diversification policy was now evident. Two dedicated freighters undertook three contracts for Royal Mail. Following the demise of Highland Airways in 2010, a newspaper freight contract as well as the Stornoway– Benbecula Public Service Obligation contract had been taken over. Two of the Saab aircraft were committed to North Sea oil and gas related charter work. A PSO contract for the Irish Government commenced in November 2011 for the provision of services between Donegal and Dublin.*

*For the year ended March 2011, a profit of £2.8 million was achieved on a turnover of £60.5 million with passenger carryings the lowest for five years at 487,460. The aircraft fleet comprised 14 Saab 340Bs; 2 Saab 340A freighters; 2 de Havilland Twin Otters and 2 Britten-Norman Islanders.*

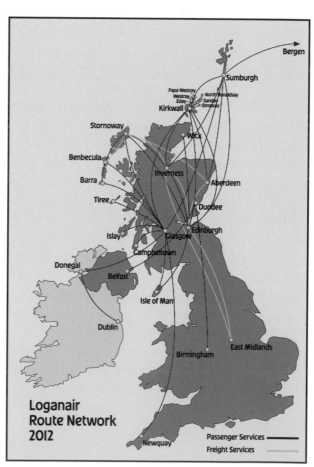

### *Where are they now?*

The airline industry has undergone enormous change in these fifty years of Loganair's existence. The majority of the main UK players, such as British Eagle, British United, British Caledonian and Dan-Air have all risen and fallen. Scotland in particular has proved not to be a happy hunting ground for airlines with at least fifty Scottish-based airlines and aviation companies having come and gone in that time. If airlines which have served Scotland were included, the number of airline casualties would be several times that figure.

Nor have Loganair's owners fared much better. Duncan Logan (Contractors) Ltd, having been a major force in Scotland's construction industry since the Second World War, never recovered from the difficulties of the Kingston Bridge contract and went into liquidation in 1975. The Royal Bank of Scotland, having dutifully borne Loganair's losses until a credible purchaser for Loganair was found in 1983, rose to global prominence until crashing in 2008 and is now eighty-two percent owned by the Government. British Midland too rose to become the UK's second biggest domestic carrier after British Airways and the second biggest operator at London Heathrow before incurring huge financial losses, becoming a wholly-owned subsidiary of Lufthansa, and ultimately suffering the indignity of being divided up and sold off.

And what of the mighty British Airways – so long Loganair's *bête-noir* in Scotland? Having served the Highlands and Islands after the Second World War with a sense of *noblesse oblige*, then adopting a policy of fighting off all-comers on small regional routes for decades when viability was manifestly unattainable, they suddenly and spectacularly relinquished interest in all Scottish routes other than those between Glasgow, Edinburgh and Aberdeen to the London airports.

And Loganair? Thanks to the unstinting efforts and commitment by innumerable staff throughout these fifty years; the Royal Bank of Scotland paternalism during the 1970s; enormous goodwill shown to the Company throughout the Scottish Highlands and Islands, and a great deal of luck, Loganair has survived to fight another day.

## Appendix I

# Not a happy hunting ground

Scotland has not been a happy hunting ground for the survival and prosperity of civil airlines and air taxi operators. Loganair was founded in 1962. In the fifty years that have passed since then, many other Scottish air transport operators have tested the market with varying degrees of success, but have ultimately disappeared, as this list, which does not claim to be exhaustive, shows.

| Airline | Date | Base |
|---|---|---|
| Aberdeen Airways | 1989-1992 | Aberdeen |
| Aberdeen London Express (ALEX) | 1994-1994 | Aberdeen |
| Ace Scotland | 1966-1966 | Glasgow |
| Air Alba | 1991-1996 | Inverness |
| Air Caledonian | 2004-2005 | Prestwick |
| Air Charter Scotland[1] | 1975-1983 | Glasgow |
| Air Ecosse | 1977-1986 | Aberdeen |
| Air Orkney | 1981-1984 | Kirkwall |
| Air-Scotland | 2003-2006 | Glasgow |
| Air Shetland | 1978-1979 | Sumburgh |
| Airgo Charter Services | 1972-1975 | Glasgow |
| Alidair Scotland | 1976-1981 | Aberdeen |
| Assistair Limited | 1974-1975 | Glasgow |
| BEA Scottish Airways | 1971-1974 | Glasgow |
| Black Isle Air Services Limited | 1961-1962 | Inverness |
| Bon Accord Airways | 1992-1995 | Aberdeen |
| Bon-Air Flight Limited | 1972-1973 | Aberdeen |
| British Airways Highland Division | 1981-1996 | Glasgow |
| British Caledonian Airways | 1970-1988 | London Gatwick |
| Burnthills Aviation | 1974-1983 | Glasgow |
| Business Air | 1987-1998 | Aberdeen |
| Cal Air | 1988-1988 | London Gatwick |
| Caledonian Airways | 1961-1970 | Prestwick |
| Capital Services Aero Limited | 1961-1962 | Edinburgh |
| Chieftain Airways | 1987-1987 | Glasgow |
| Cirrus Aviation Limited | 1962-1979 | Strathallan |

| | | |
|---|---|---|
| City Star Airlines | 2005-2008 | Aberdeen |
| Donaldson International Airways | 1969-1974 | London Gatwick |
| Eastern Seaboard Limited | 1972-1973 | Dundee |
| Edinburgh Air Charter | 1998-1999 | Edinburgh |
| Edinburgh Flying Services | 1971-1979 | Edinburgh |
| Euroscot Express | 1997-1999 | Bournemouth |
| Flyglobespan | 2002-2009 | Edinburgh |
| FlyWhoosh | 2007-2007 | Dundee |
| Frasair | 1982-1983 | Glasgow |
| Grampian Air Taxies Limited | 1972-1973 | Aberdeen |
| Highland Airways | 1996-2010 | Inverness |
| Highland Express | 1987-1987 | Prestwick |
| Lakeside Aviation | 1991-1993 | Aberdeen |
| Macair | 1995-1995 | Edinburgh |
| Malinair | 1985-1987 | Glasgow |
| McDonald Aviation Limited | 1974-1976 | Dundee |
| Merlin Executive Aviation | 1991-2001 | Glasgow |
| Peregrine Air Services | 1969-1989 | Inverness |
| Robertson Aviation Limited | 1975-1976 | Glasgow |
| Scottish European Airways | 1988-1990 | Glasgow |
| Site Aviation | 1973-1974 | Aberdeen |
| Strathallan Air Services (Strathair) | 1962-1971 | Strathallan |
| Tartan Air Charter | 1976-1977 | Glasgow |
| West Highland Aviation | 1986-1988 | Skyc |

[1] Air Charter Scotland along with the fixed-wing operations of Burnthills Aviation, maintenance company Strathclyde Aero Services, Frasair, aircraft sales data company Computaplane, and consultancy Duncan McIntosh Aviation, combined in 1983 to form Air Charter (Scotland) Ltd.

## Appendix II

# Loganair's People

**Company Personnel and date of takeover by Commercial Bank – October 1968**

**GLASGOW**
Duncan McIntosh – Managing Director
Kenneth Foster – Chief Pilot/Training Captain
Geoffrey Rosenbloom – Deputy Chief Pilot
John Grindon – Line Pilot
Bill Henley – Line Pilot
Maurice Barron – Line Pilot
Ann Chalmers - Office Supervisor
Stuart Lock – Operations Officer
Elaine Gilbert – Typist
Agnes Anderson – Typist
Jessica Rae – Typist
Marion McGregor – Office Cleaner
Walter Ramsay – Chief Engineer
James Miller – Deputy Chief Engineer

Gilbert Fraser – Hangar Foreman
Kenneth Spratt – Engineer
Albert Tester – Engineer
George Cormack – Engineer
Kenneth Milligan – Engineer
Alexander Haddow – Engineer
James Jackson – Engineer
David McKenzie – Apprentice Engineer

**ORKNEY**
James Lee – Jt. Senior Pilot, Orkney
Andrew Alsop – Jt. Senior Pilot, Orkney
Helen Manson – Office Supervisor
Michael McTurk – Engineer
Bryan Sutherland – Apprentice Engineer

**Loganair Staff at 31 January 2012**

**GLASGOW**
**Pilots**
Graeme Abernethy
Fraser Beaton
Trudy Johnston
Harry Lockhart
Lionel McClean
Mairi Nicholson
Samantha Pollak
**Pilots – Admin**
Elaine Harvey
Tim Kinvig
Janette MacArthur
David Miller
Sandra Morison
Andrew Robertson
David Rubery
Stephen White
**Pilots - 340 Fleet**
Michael Anderson
Annag Bagley
Ian Bottomley
Alex Brand
Graham Bunney
Sophia Chowdhury
Stephen Clark
Donna Clayton
Jonathan Crawford
Paul Deakin
James Donaldson
Ashley Eagles
Peter Finnie
Andrew Forey
Alistair Forrest
Robin Freeman
Murray Gibbons

Justin Gore
Ian Hammond
Jakob Hansen
Lars Harboesgaard
Jeremy Hawksworth
Sarah Hendry
Neil Hughes
Ryan Jones
Greg Logan
Barry Lyall
James MacDonald
Andrea Marco
Mark Millar
Christopher Miller
Nathan Parker
Matthew Payne
Mark Pendlebury
Christopher Rennie
Kenneth Roberts
Rebecca Simpson
Lucy Sinclair
Bryon Smee
Graeme Smith
Samantha
Walkinshaw
Paul Wickman
**Cabin Crew**
Margaret Affleck
Jill Armour
Linda Clayton
Tracy Colvin
Claire Dare
Natalie Donaldson
Steven Duggan
Sheena Elphick
Lissa Forrest

Pamela Greenlees
Gillian Hunter
Christina Johnstone
Margaret Kinvig
Helen Masters
Tracy Morrison
Laura Nelis
Andrea O'Donnell
Geraldine Parker
Roseleen Rawdon-
Leitch
Kenneth Revis
Julie Robertson
Paul Seery
Lynne Smith
Patricia Thistle
Fiona Thomson
Martyn Tulloch
Helen Walker
Sheree Wallace
**Cabin Crew -**
**Management**
Avril McEwan
Eilidh MacMillan
**Engineering - Line**
Adam Currie
Greg Grindlay
Alfred Holt
Greig Hume
Stuart Kay
Nathan Leitch
Mark McEwan
Alistair Miller
Stuart Murdoch
Edwin Muzaale

James Muzaale
Charles Ross
Philip Smith
Peter Welch
Iain Welsh
Andrew Young
Shingirai Zimuto
**Line Support**
Keith Morbin
Abdulkarim Nathani
Neil Ritchie
Craig Stewart
**Engineering**
**Training**
Peter Bainbridge
**Engineering - Base**
Stuart Binning
Thomas Boyd
Richard Brown
Robert Campbell
Iain Davidson
Ross Gourdie
Alex Harley
Derek Johnstone
Michael Kaloheris
Gerald Kerr
Stuart Martin
John McArthur
Steven McDade
Paul McDonald
Kelly McKinstray
Mark Moraetes
Harry Nodoro
Thomas O'Donnell
Tom Ritchie

James Robertson
David Royle
James Russell
Ross Simpson
James Stevenson
Sean Stewart
Neil Swan
Suresh Uppalapati
James Wardrope
Neil Weppenaar
Kyle White
Andrew Wilson
**Engineering Stores**
Stewart Brown
Robert Currie
Nicola Dick
Reginald Dunlop
Gail Hancy
John Kidd
Hugh Lyden
Iain McDade
Alastair Ness
Andrew Paterson
Archibald Thomson
Elaine Walmsley
**Engineering**
**Workshops**
Kenneth Macleod
Philip Morhulec
**Technical Records**
Alan McCartney
Peggy McCartney
Michelle
McCorrisken
Douglas Milne

**Technical Services**
Stuart Alexander
Michael Graham
Geoffrey Provan
Emma Rae

**Operations**
Jacqui Aubrey
Kay Brown
Lorna Connell
Gavin Irving
Sandra Jones
Martin McWilliam
Janette Shiels
Neville Walmsley
Suzanne Wright

**Administration - CEO**
James Adams
Lloyd Cromwell Griffiths
Linda Doak
Scott Grier
Roger Hage
David Harrison
Jonathan Hinkles
Robert McLellan
Stuart McMahon
Brian Mitchell
David Ross

**Administration**
Valerie Gillespie
Ruth Glendinning
Sarah Gunn
Marion MacKay
Claire McCafferty
Heather McLean
Fiona Tennant

**Human Resources**
Shirley Orr
Audrey Sanders

**Quality**
Graeme Anderson
Lorna Cowper
Jennifer Cuthbertson
Brian Robertson

**Revenue**
Roy Bogle
Richard Craig
Iain Hunter
Colin Munro
Ian Ritchie
Donna Sutherland

**IT**
Jennifer Berrie
Stephen Buckley

**Customer Services**
Graham Everett

Clare McDade
Shirley Nicolson

**EDINBURGH**
**Pilots**
Paul Blackler
Frances Devlin
Stuart Dickson
Euan Downie
Stewart Easson
Christopher Hammersley
Thomas Holloway
Gordon Hutchison
Christopher Lamb
Gordon MacFarlane
Hunter Mercer
Ronan Milne
Fergal O'Neill
Gareth Pritchard
Almudena Rivas
Malcolm Sinclair
Edward Watt
Richard Westbrooke
Russell Wheatley

**Cabin Crew**
Deborah Boyle
Leanne Corrigan
Jessica Fuentes
Gillian Gibson
Kirsty Gilmour
Rachel Gilmour
Catherine Lund
Jody MacRae
Charmaine Hagan
Margaret McCallum
Joanne Meaney
Sara Parfitt
Karen Todd
Kirsty Walter

**Engineering**
George McInnes
David Muir
George Peffers
Michael Stanley

**ABERDEEN**
**Pilots**
Richard Anderson
Thomas Attrill
Douglas Colman
Charles Coulson
Eric Dawson
Amelia Findlater
Sean Garswood
Benjamin Hyatt
Petter Krantz
Christopher Lodge
Simon Merritt
Steven Murphy

Tom Myles
James Parker
Duncan Peace
John Ratcliffe
Erlend Reklev
Gary Rumbles
David Smith
Jonathan Smith
Martin West
James Workman

**Cabin Crew**
Jennifer Anderson
Nichola Charlish
Iain Cowe
Angela Fraser
Clare Grant
Marian Scott
Angela Shewan
Claire Thomson

**Engineering**
Eric Baillie
William Bruce
Neil Butler
John Douglas
Louis Fernando
David Martin
Ian Martin
Mark Parrish
Roderick Smith

**INVERNESS**
**Pilots**
Johan Alf
Robert Blackburn
Alex Carolan
Tim Curwen
Sean Darwent
Dawn Hunter
Scott Lindsay
Scott Martin
Alistair Mclean
James Morris
Lars Ojeskog
Euan Patrick
Nigel Pybus
Craig Sharkey
James Slater
Marc Stewart
William Temple

**Cabin Crew**
Lisa Bain
Alistair Grant
Faith MacIntyre
Megan McBean
Elizabeth Milne

**Engineering**
Carl Cheeseman
Thomas Eaton
David Fox

John Hawick
Gregory Sparrow

**DUNDEE**
**Pilots**
Peter Anderson
Justin Austin
Michael Cryle
Alexander Torrance
Kevin Utting
Stewart Webb

**Cabin Crew**
Laura Burnett
Dana Taylor
Stacey Turner

**Engineering**
Geoffrey Sloan
Simon Watkins

**KIRKWALL**
**Pilots**
John Bain
Malcolm Hempsell
Stuart Linklater
Colin McAllister

**Engineering**
Stephen Cogle
Jamie MacKenzie
Alan McCaffrey
Victor Murphy
Kenneth Ross
Paul Schinkel
Bryan Sutherland
Thorfinn Thomson

**Customer Services**
Michael Bain
Inga Bosworth
Susanne Carter
Kenneth Christie
Jacqueline Delaney
Mark Drever
Richard Falconer
Premysl Fojtu
Richard Gorn
Michael McManus
Jennifer Nichol
Alex Rendall
Deenesh Revis
Lyndsay Robinson
John Smith
Charlotte Stanger
Kenneth Stevenson
Calum Walter
Ryan Walter
Gemma Wilson

**STORNOWAY**
**Customer Services**
Keith Barron
David Brown
Marina Campbell

Jeffrey Fairclough
Brian Farrell
Garry Finlayson
William Innes
Iain MacIver
Andrew MacLeod
Iain MacLeod
Ivor Macleod
Martyn MacLeod
Roderick MacLeod
John MacRae
Darren Salter
Jillian Scott

**ISLAY**
**Customer Services**
Margaret Anne Dumigan
Pamela Ferguson
Stacey Green
Susan Hamilton
Claire Logan
Kirsten MacIntyre
Karen McNiven

**TIREE**
**Customer Services**
Carol Ann MacArthur
Mairi MacArthur
Isabella MacKinnon

**WICK**
**Customer Services**
Paula Duffy
Helen Fraser
Annette Gow
Jacqui Gow
Maureen Henderson

**BENBECULA**
**Customer Services**
Iain Bagley
David Lister
Rhoda MacCormick
Neil MacDonald
John MacKay
Colin MacLeod

**BARRA**
**Customer Services**
Janet MacLean
Naomi Robarts
Elaine Stewart

**CAMPBELTOWN**
**Customer Services**
Jane McCallum
Lesley McLean
Margaret McSporran

# Bibliography

***Secondary Sources***

Calderwood, Roy, *Times Subject to Tides: The Story of Barra Airport* (Erskine: Kea Publishing, 1999)

Cameron, Dugald, Glasgow's Airport (Edinburgh: Holmes McDougall, 1990)

Cameron, Dugald, *A Sense of Place: Glasgow Airport at the Millennium* (Paisley: BAA Glasgow/Squadron Prints, 2000)

Cameron, Dugald and Roderick Galbraith, *A Hundred Years of Aviation in Scotland* (Glasgow: Foulis Press, 1996)

Fresson OBE, Captain E E, *Air Road to the Isles*, 2nd Edition, (Erskine: Kea Publishing, 2008)

Gunston, Bill, *Diamond flight: The story of British Midland* (London: Henry Melland, 1988)

Guy Halford-MacLeod, Guy, Britain's Airlines: Vol. 3, 1964 to Deregulation, (Stroud: The History Press, 2010)

Hutchison, Iain, *The story of Loganair: Scotland's airline – The first 25 years* (Stornoway: Western Isles Publishing, 1987)

Hutchison, Iain, *Air Ambulance: Six Decades of the Scottish Air Ambulance Service* (Erskine: Kea Publishing, 1996)

Hutchison, Iain, 'The Scottish Air Ambulance Service, 1928-1948', *Journal of Transport History*, 2009, 30:1, 58-77.

Jones, A C Merton, *British Independent Airlines, 1946-1976* (West Drayton: The Aviation Hobby Shop, 2000)

Lo Bao,Phil, *An Illustrated History of British European Airways* (Feltham: Browcom, 1989)

Lo Bao, Phil and Iain Hutchison, *BEAline to the Islands: The story of air services to offshore communities of the British Isles by British European Airways, its predecessors and successors* (Erskine: Kea Publishing, 2002)

Merton Jones, A C, *British Independent Airlines 1946–1976* (West Drayton: TAHS, 2000)

Warner, Guy, *Orkney by Air: A photographic journey through time* (Erskine: Kea Publishing, 2005)

Whitfield, Alan, *Island Pilot*, 2nd Edition (Erskine: Kea Publishing, 2007)

Woodley, Charles, *Scotland's Airlines* (Stroud: The History Press, 2008)

## Primary Sources

Aviation and Travel Consultancy Ltd - *An Expanded Air Service Network for the Highlands and Islands*, 2003

Civil Aviation Authority – Decisions on Air Transport Licence Applications

Civil Aviation Authority – *Air Transport in the Scottish Highlands & Islands*, 1974

Department of Transport - *The Future Development of Air Transport in the United Kingdom: Scotland*, 2002

Highlands & Islands Development Board – Annual Reports

House of Commons Expenditure Committee (Trade & Industry Sub-Committee) Reports

House of Commons Transport Committee - *Aviation - Sixth Report of Session 2002 - 2003*

Loganair Annual Reports & Accounts 1967–2011

Loganair Board Minutes 1966–2011

Mott MacDonald Ltd - *A Review of Air Services in the Highlands and Islands*, 2009

Report of the Highland Transport Board – Highland Transport Services, 1967

## Periodicals

Glasgow Herald

Holyrood Magazine

Oban Times

Orcadian

Press & Journal

Scotsman

Scottish Air News

Scottish Transport Review

Shetland Times

STAN – Scottish Travel Agents News

Stornoway Gazette

West Highland Free Press

# Index